CHARCUTERIE

Also by Michael Ruhlman and Brian Polcyn

SALUMI: THE CRAFT OF ITALIAN DRY CURING

Other food books by Michael Ruhlman

THE MAKING OF A CHEF: MASTERING HEAT AT THE CULINARY INSTITUTE OF AMERICA

THE FRENCH LAUNDRY COOKBOOK *(WITH THOMAS KELLER AND SUSIE HELLER)*

THE SOUL OF A CHEF: THE PURSUIT OF PERFECTION

A RETURN TO COOKING *(WITH ERIC RIPERT)*

BOUCHON *(WITH THOMAS KELLER, JEFFREY CERCIELLO, AND SUSIE HELLER)*

THE REACH OF A CHEF: PROFESSIONAL COOKING IN THE AGE OF CELEBRITY

UNDER PRESSURE: COOKING SOUS VIDE *(WITH THOMAS KELLER, SUSIE HELLER, AMY VOGLER, JONATHAN BENNO, AND CORY LEE)*

THE ELEMENTS OF COOKING: TRANSLATING THE CHEF'S CRAFT FOR EVERY KITCHEN

RATIO: THE SIMPLE CODES BEHIND THE CRAFT OF EVERYDAY COOKING

LIVE TO COOK *(WITH MICHAEL SYMON)*

AD HOC AT HOME *(WITH THOMAS KELLER, DAVE CRUZ, SUSIE HELLER, AND AMY VOGLER)*

RUHLMAN'S TWENTY: 20 TECHNIQUES, 100 RECIPES, A COOK'S MANIFESTO

THE BOOK OF SCHMALTZ: A LOVE SONG TO A FORGOTTEN FAT

CHARCUTERIE

The Craft of Salting, Smoking, and Curing

. . . . .

Revised and Updated

. . . . .

Michael Ruhlman
AND Brian Polcyn

Illustrations by Yevgeniy Solovyev

W. W. Norton & Company NEW YORK LONDON

For information about permission to reproduce selections from this book,
write to Permissions, W. W. Norton & Company, Inc.,
500 Fifth Avenue, New York, NY 10110

For information about special discounts for bulk purchases, please contact
W. W. Norton Special Sales at specialsales@wwnorton.com or 800-233-4830

Manufacturing by Courier, Kendallville
Book design by Barbara Bachman
Production managers: Andrew Marasia, Devon Zahn

Library of Congress Cataloging-in-Publication Data

Ruhlman, Michael, 1963–
Charcuterie : the craft of salting, smoking, and curing / Michael Ruhlman
and Brian Polcyn ; illustrations by Yevgeniy Solovyev. – Revised and
updated.
pages cm
Includes index.
ISBN 978-0-393-24005-4 (hardcover)
1. Smoked meat. 2. Food–Preservation. 3. Cooking (Smoked foods)
I. Polcyn, Brian. II. Title.
TX609.R84 2013
641.6′16–dc23
2013019905

W. W. Norton & Company, Inc.
500 Fifth Avenue, New York, N.Y. 10110
www.wwnorton.com

W. W. Norton & Company Ltd.
Castle House, 75/76 Wells Street, London W1T 3QT

1 2 3 4 5 6 7 8 9 0

For Julia and Donna,
Alana, Alex, Carmen, Dylan, Ben, Addison, and James

CONTENTS

Charcuterie is everywhere around us, but most in America don't recognize it as such. Bacon, sausages, hams, pâtés, and terrines are all part of this great culinary specialty. In the world of cooking, charcuterie is in a class by itself.

My first exposure to charcuterie, like so many Americans, was cold cuts: Oscar Mayer bologna and salami. Unless you were the child of European immigrants, you probably never had a great dry-cured sausage, a saucisson sec, or a soppressata as a kid. When I was growing up in Florida in the 1970s, charcuterie like that wasn't available. Thanks to the extraordinary changes in this country, you can now find it in supermarket chains.

Americans travel more than ever and are more likely to explore regional specialties throughout the world. As we move through our lives, as we travel and explore, our reference points change. Our experiences of charcuterie gather and we learn that those bologna cold cuts we took for granted as kids have their roots in mortadella and the other emulsified sausages popularized in Bologna; that packaged grocery store salami is a descendant of the dry-cured sausages called *salami*, works of great craftsmanship and great flavor.

Making those food connections, and recognizing those reference points, is important because it deepens the experience of cooking and eating. And understanding that the historical roots of charcuterie reach hundreds of years back, and that the fundamental methods of charcuterie, namely curing and preserving, reach all the way back to earliest civilization, makes us realize that this specialty is one of the most important kinds of cooking there is.

In this book, Michael Ruhlman and Brian Polcyn subdivide charcuterie into its component parts, first describing curing meats, fish, and vegetables with salt. They explore the many variations of the pâté and the versatility of the confit. But of all the food we call charcuterie, my favorite by far is the sausage, and sausage is the backbone of *Charcuterie*. What's best about this part of the book is that Michael and Brian not only describe in detail the various stages of sausage making but also isolate the key steps and techniques that can elevate a good sausage to a great one. In addition, they give what amounts to a primer in the key points of dry-curing sausages and whole cuts of meat, a segment of charcuterie that's really in its infancy here in America.

This book is important because so few people understand what charcuterie is or recognize it as the great branch of cooking that it is. Charcuterie represents some of the oldest methods of cooking, and so has deep culinary roots and an important role in the development of civilization. It has a long and varied place in restaurant kitchens, and is now enjoying something of renaissance in the American restaurant, especially with regard to dry-cured items. Charcuterie is amply represented at my bistro, Bouchon, and is also at home at the French Laundry and Per Se, my four-star restaurants.

At Per Se, Joshua Schwartz makes a form of lardo, salting and dry-curing fatback from just above the pork shoulder, and serves very very thin slices of it. For a special dish, he'll wrap a slice of lardo around an asparagus tip and a slice of truffle and serve it as a canapé. But lardo is delicious on its own. Its texture and flavor are amazing–there's nothing like it–and it couldn't be easier to make. All it takes is good pork–that is, organic or farm-raised pork–which is now available not just to restaurant chefs but to home cooks everywhere via the Internet. I want to stress how simple and natural the method is: good pork, salt, and time are the principal ingredients. That something so easy to make is also so extraordinary to eat is part of its pleasure.

Charcuterie is appropriate at every level of dining, and it runs the entire gamut of cooking. That includes home cooking, where charcuterie once played a huge role and will again–not one day far in the future but now. Some charcuterie techniques couldn't be simpler. A BLT is one of the best uses of charcuterie I know, for instance. Making your own sausage and frying up some patties is no more difficult than grinding your own hamburger, but to season it yourself, and to cook and eat what you've made, is a very special thing. Charcuterie is sometimes relegated to fall and winter cooking, but it should be an important part of your kitchen year-round.

A final reason *Charcuterie* is important: it recognizes the pig as the superior creature that it is. From a culinary standpoint, the pig is unmatched in the diversity of flavors and textures it offers the cook and the uses it can be put to–from head to tail, from ham to tenderloin, it's a marvel. A piece of pork belly can be brined, roasted, grilled, sautéed, dry-cured, braised, or confited, with widely varying results. This is a very hopeful time for the pig in America, and this book underscores that fact.

Come to think of it, this book reminds me what a hopeful time it is for cooking in general in this country. This may well be the most exciting time ever to be a cook and a chef in America. And *Charcuterie* is a perfect example of why.

—**Thomas Keller**
Fall 2005

INTRODUCTION TO THE REVISED EDITION

A book devoted to animal fat and salt, one comprising recipes that can take days and even months, was not expected to sell beyond a small, slightly disturbed band of committed home cooks and chefs, especially in a country, America, that had all but waged war on animal fat and salt and instructed its citizens to avoid both at all costs.

So it came as a happy surprise to our publisher that, when it was first published in the fall of 2005, it did okay, not bad even. But then it did the unexpected. It started doing better. Word of mouth spread. Chefs embraced the book and began offering more and more charcuterie on their menus.

Americans gravitated to the heavy use of smoke and fire and sausage making contained in the book. Cathy Barrow of Washington, D.C., and Kim Foster of New York City created a small blogging phenomenon when they began using the hashtag #charcutepalooza on Twitter, gathering more than four hundred bloggers and non-bloggers to join in a monthly charcuterie challenge and share their results.

The book's success has led many people to call it "the bible" of charcuterie, which we're flattered by, but it's a compliment that is misleading. *Charcuterie* is not a bible in that it's not exhaustive in its coverage, and as we said at the outset, the craft itself is one you practice, a craft that you always get better at, can always learn more about. The subject, in fact, is inexhaustible.

Our hunch is that cooks and chefs responded to the lively curiosity and expressiveness of the book, its joy in its subject, its headlong embrace of fat, and the way we were able to clarify and simplify what has always seemed complex and difficult.

Because the book continues to sell, continues to be embraced by incoming waves of young culinarians, we wanted to amend the book as we've continued to deepen our knowledge of the subject. We've taken out ingredients that were used in commercial sausage production but that aren't needed at home, and we've adjusted salt levels where necessary. (Salt preferences vary; salt to your own tastes; we like 1.75 percent salt in our sausages, but you may want 1.5 percent–there are no hard and fast rules in the kitchen, though we hasten to add that the most important of your senses is your common sense, and this should be used at all times, rather than strict adherence to rules and recipes.)

We've gone on to explore the Italian craft of dry curing in a book called *Salumi*, because of increasing interest in it. When we first wrote *Charcuterie*, there was very little out there in the way of explaining products such as mortadella and soppressata, and so we included these Italian preparations in a book devoted to French preparations. We've retained most of these, excepting guanciale, which we feel belongs exclusively in *Salumi*, especially given that French charcutiers don't commonly cure the jowl. In any case, this is why you'll still see some preparations here that lie outside the charcutier's purview.

The craft of charcuterie has been practiced for many centuries if not millennia, but it was long absent from American kitchens. We are thrilled to be a part of its resurgence.

—**Michael Ruhlman and**
Brian Polcyn

CHARCUTERIE

1

THE REASON FOR THIS FOOD, THIS BOOK:
WHY WE STILL LOVE AND NEED HAND-PRESERVED FOODS
IN THE AGE OF THE REFRIGERATOR, THE FROZEN DINNER,
DOMINO'S PIZZA, AND THE 24-HOUR GROCERY STORE

. . . . .

Sometimes books are the result of a surprise, in this case a surprise (via duck confit) that became a fascination that transformed into a quest to understand this food that we still categorize under the broad label of charcuterie, a range of preparations from sausages to pâtés, confits to cured salmon, all of which have some sort of cure and preservation at their core. This is not a thirty-minute-meals cookbook, not a book to help you get dinner on the table fast or tell you how to whip together an impromptu dinner party for eight. It is a book about craftsmanship, for people who love to cook and eat. In this chapter, Brian and I describe its genesis—that is, why on earth we would devote two years of our lives and an entire cookbook to a love song to animal fat, to salt, to the pig—as well as how to use our recipes.

A powerful mania descended on me a decade ago when I first tasted duck confit (*confit de canard*): duck salted for hours, if not days, then poached gently in its own fat, and then submerged in that fat and left to "ripen." What amazed me first was that you could poach meat in fat. I found the idea of poaching anything in fat appealing, and the idea of poaching a rich fatty meat in more fat enormously so. But what truly hooked me was how amazing the duck confit was to eat—the salty, gently spiced meat was deeply, richly succulent, the skin crispy.

I began to explore the technique. You could do so many things with duck confit: stew it in some beans (as in cassoulet, the well-known French bean dish), make a salad of it with some greens and a vinaigrette, put it on mashed potatoes or polenta, add chunks of it to a country pâté. You could even tuck it between slices of Wonder Bread and it would still be fantastic.

But the ideal way to serve it, I realized, is on a bed of diced potatoes that have been fried crispy and flavorful in that amazing duck fat. Duck served this way has what chefs call integrity: both items are cooked in the same medium, with different effects, and the dish retains the rustic simplicity embodied by the duck and the potato. This is what you would eat on a farm in France, where the technique of preserving duck has thrived for centuries.

Yet originally duck that pleased the palate and satisfied the soul wasn't the point. The goal of the confit was preservation; that's what *confit* means: preserved. A French farmwife cooked and stored duck in its own fat because that kept the duck from spoiling for a long, long time. She and her family, frugal by necessity, could eat it as they needed it, wasting nothing. That pleasure happened to be a by-product of economy and survival underscored the ingenuity of the method.

As my obsession grew, I made and ate a lot of duck confit. I annoyed my wife at restaurants by demanding to know from our waiter the exact method, type of cures, type of fat the chef used in the confit (one chef, for instance, used half pork fat because he liked the flavor). But my appreciation for it didn't peak until its greatness dawned on me: preservation techniques, while no longer necessary today, could result in astonishing food.

I asked a chef friend, a teacher expert in the ways of preservation, Dan Hugelier, why now, given that we can "preserve" food fine in a fridge or freezer or in Cryovac, sealed in oxygenless packages, why was confit, why was charcuterie–a culinary specialty largely defined by preservation methods–still relevant? Dan looked at me as if I were an idiot and said, "Taste." Having solved the survival issue, we have the luxury to think about pleasure, about refinement.

You can confit many cuts of meat. Goose, in addition to duck, of course, chicken or turkey–or a pork loin. It's a remarkable thing: you can buy a supermarket pork loin, unnaturally lean now and as flavorful as cardboard, and, with the basic confit method, turn it into something so tasty you'd swear voodoo were involved. You can also turn pork shoulder into a mean confit. At a restaurant in Seattle, I once ordered deep-fried pork belly confit–more or less a chunk of deep-fried fat–and I nearly fell over backward it was so good, crisp on the outside, melty and spicy inside (see page 265 for the recipe).

A pâté, a way of alchemizing scraps into culinary treasure, is another form of food preservation. As are sausages, bacon, ham, smoked salmon, smoked trout, or simple lox, salmon cured with salt and seasonings. All these items are part of the specialty called charcuterie, and each grew out of the need for preservation. Contemporary chefs have adopted some of these preservative techniques for foods you might never think of trying to preserve. Halibut confit would sound ridiculous to a French farmer, but prepared carefully, it's delicious.

Derived from the French words for flesh (*chair*) and cooked (*cuit*), the term *charcuterie* came to designate the shops in fifteenth-century France that sold products of the pig including offal. The Romans, who made standards of raising, killing, and cooking of pork points of law, regulating its production, were likely the first to turn pork butchery into a trade, but it was the French charcutier who brought the greatest ingenuity to pig preparations. In the fifteenth century, charcutiers were not allowed to sell uncooked pork (though they could sell uncooked fat, which would be rendered into lard at home and used for cooking there), and so they created all manner of cooked (or salted and dried) dishes to be sold later—pâtés, rillettes, sausages, bacon, trotters, head cheese. The charcutiers of the late fifteenth century, the time when the first guilds were formed, were highly esteemed. These tradesmen in charge of pork butchering played a critical role in maintaining the food supply in their towns; charcuterie then meant cooking and preserving the meat for a community. Long before the Renaissance, and through the Industrial Age, societies, civilization depended on such preservative techniques. By the time of the French Revolution, nearly one hundred master charcutiers were plying their trade in the country's capital.

The history of charcuterie, in the sense of salting, smoking, and cooking to preserve, may date almost to the origins of *Homo sapiens*. It has been carried on in many forms through virtually every culture, and it has been one of the foundations of human survival in that it allowed societies to maintain a food surplus and therefore helped turn early peoples from nomads into clusters of homebodies. Sausage recipes date to before the golden age of ancient Greece. Even before that, the Egyptians were fattening geese for their livers—and possibly making the first pâté de foie gras.

In fact, the need to preserve food may well have been what led us to cook it in the first place, and then only by accident. It's not unlikely that the ancestors of *Homo sapiens* hung surplus raw food over a fire to keep away bugs and animals. In the morning, they discovered that it was smoked hot, tender, and delicious.

Other historians have suggested that our ancestors first discovered cooked food in the form of animals that had perished in forest fires, and then began to cook food on purpose. Regardless of how they discovered cooking, they surely realized that cooking made food not only taste good but last longer as well.

At about the time my fascination with confit, and then preservation techniques generally, plateaued, I met Brian Polcyn, chef of Five Lakes Grill in Milford, Michigan, about forty-five miles west of Detroit. I'd just finished a year's study at the Culinary Institute of America in Hyde Park, New York, in order to write a book about how one learns to cook professionally. I was green, but I'd learned the core cooking fundamentals and was eager to know more about the profession. I finagled a magazine assignment that allowed me to return to the CIA to observe the Certified Master Chef Exam, a ten-day marathon of cooking in all kinds of styles–classical French, traditional Asian, regional American, nutritional, patisserie–all fiercely graded. There was something insane about the test and those who took it, but by then I'd learned that insanity can sometimes more than adequately describe a chef. Imagine performing ten Iron Chef competitions back to back every day for ten days. The test is given by the American Culinary Federation at the Culinary Institute; that particular test in the spring of 1997 included a total of seven chefs, Brian among them.

Covering the test, I gravitated toward and spent the most time with Brian because he was unfailingly articulate and so he made for both good copy and a good education; also, as he was on the verge of passing until the very end, his story was dramatic. Incidentally, he was expert in the specialty called charcuterie.

The judges graded the participants' pâtés and terrines, the backbone of charcuterie, on two separate sections of the test, but mastery of charcuterie principles was useful throughout the exam. What to do with a pound of scallops, for example, in the mystery basket–a form of cooking test in which chefs must cook a complete meal using a tricky, limited selection of ingredients–but make a mousseline? Then perhaps use that rabbit trim and pork shoulder to make a sausage? This was not by accident. It's not difficult to roast a rack of lamb and make it taste good, but to transform scraps into delicious food requires craftsmanship and knowledge–that is a true measure of a chef. An ability to make delicate, lovely use of the roughest, strongest, least desirable parts of an animal describes the overall excellence of a chef.

I wrote about Brian taking the test in an article and in a subsequent book, *Soul of a Chef*, and we became friends. In the meantime, soon after he returned home from the test, he was asked by Schoolcraft College, in Livonia, Michigan, outside Detroit and not far from his Five Lakes Grill, to become a chef-instructor in, coincidentally, charcuterie.

It's in the nature of the chef to accept with cheerful willingness a workload that is completely impossible. It's a matter of pride and a personal challenge, even, taking on a third job when two full-time jobs are really already just a little too much. And a chef does this not wincingly or with a sigh, but rather with a breezy immediate response: Sure.

Brian's days began at 5:30. He taught and lectured on butchery in the morning, "demo-ing" whole pigs and lambs, taking primal cuts of beef down to the subprimals of shoulder, shank, strip loin, tenderloin, blade, flank. He taught charcuterie in the afternoon. Then school let out around three, at which time he would cruise a half hour to Five Lakes Grill, where as owner he still worked the line. He might have to leave the line briefly when one or another of his five kids had a soccer game—as coach, he had to be there—or to interrogate his teenaged daughter's suitors, but the majority of his waking hours were spent making, serving, or teaching food and cooking.

By the time a critic from *The Atlantic Monthly* showed up to review his restaurant, Brian had been preparing charcuterie for two decades and teaching the art and craft of it for several years. So it's not surprising that while his restaurant offers an array of contemporary American regional cuisine, it was the pork, duck, and pâté that Corby Kummer singled out:

> The pork and the duck were the best I've had in years—anywhere, even in south-western France, where every house is a farm and every farm fattens a few ducks. Specifically, Polcyn's forte is charcuterie. . . . Every day a different pâté or ter-rine is offered, and the peppery duck pâté I tasted was a tour de force. Each component—the firm little chunks of duck leg, the pistachios, a soft pink-and-red forcemeat of pork and duck—had distinct texture and flavor; the aftertaste was clear and pleasant, with none of the muddy residue most pâtés leave.

Brian is an exuberant chef, devoted family man, and articulate teacher, but, most impor-tant, he loves everything he does and, in the culinary realm, he especially loves charcuterie. We'd remained in touch, and when I called him to say I wanted to write a book about this sub-ject that I too love, he happily said, "You bet."

Brian, a native of Michigan, born to a Polish father and Mexican mother, has a lively manner and a teaching style that wouldn't be out of place on the comedy circuit. ("This week-end at the restaurant," Brian will tell his students to begin the day's lecture, "I smoked some duck breast. Excellent flavor, but really hard to keep lit.") He grows so excited when speaking about food and cooking he bounces as he talks.

"My Polish grandma, my father's mother, made kielbasa every Christmas and Easter," Brian told me when we sat down to talk about the book. "Then my mom took over the job.

Looking back on it, it was always good food, everything made from scratch. We didn't have any money, so everything was used and used well. Kielbasa was the holiday ritual. We'd grind the meat and season it. The next day we'd stuff it, tie it into big rings, hang the rings over a broom handle on chairs, put the dog out, and set the kielbasa in front of the fire overnight.

"But here's the thing. No one's been able to reproduce Grandma's kielbasa. After she seasoned the meat, Grandma would put it beneath her bed. I don't know if she was trying to keep it from us, from kitchen mice, if she thought something about the conditions under the bed were special, or if it was just superstition, but she always placed the kielbasa under the bed for one night. My mom tried to perfect it after Grandma died, working from my father's memory, but it was no use. After decades of practicing charcuterie, it's still a mystery to me.

"'Practicing charcuterie' is the right way to phrase it for two different but related reasons. The first and most important is that you're always learning, always practicing, never perfecting, because the conditions are always changing on you. Much of charcuterie is in your control, but much isn't–the humidity, the water content of the fat you're using; whether it is hot summer or chilly fall; how hot your grinder got while grinding–those things and more come into play, so that for me, I always feel that I'm practicing, always learning. Also, the work of the charcutier is like that of a doctor, who is always learning, always discovering something new about patients and treatments and care.

"For me, charcuterie is the most beautiful part of the kitchen, the most satisfying work there is. Its rich history, its diverse cultural variations, and the delicious results of these techniques, some of them as old as humankind, that's what does it for me.

"My love of charcuterie has only ripened as I've grown as a cook and a chef. When I was a kid, sausage was common on our table. I did the grinding and enjoyed it.

"But this romance with charcuterie didn't start till my early twenties, when I went to work at the Golden Mushroom, outside Detroit, for Milos Cihelka, a Czech immigrant, my mentor and one of Michigan's great chefs. I'd been cooking five or six years and thought I knew something. But when I started grinding meat and smoking sausages for Chef Milos, I realized how little I knew. I was also fascinated by the process. I got to work early and left late. I took notes like crazy. I smoked with all kinds of wood, anything I could find, apple, cherry, alder, ash, hickory, as well as nonwoods, herb branches and corncobs (corncobs make good pipes, but their smoke is awful).

"I loved smoke, this way of imparting flavor that few other chefs my age were doing. A decade later, when I became a partner in a restaurant in Pontiac, Michigan, I built a smoker in the alley behind the restaurant to smoke fish, game, poultry, sausages, and those became some of the most popular items on the menu.

"So first with Milos, and then on my own, I learned the value of taking inexpensive cuts of meat and changing the texture and flavor in a way that was very personal, completely my own. By curing, smoking, and brining my own meats, by making pâtés and terrines and mousselines, I distinguished my food from that of other restaurants in my area, and did so in a way that was unusual in Michigan and also very economical.

"I've always had some kind of smoked or cured meat, some kind of pâté or terrine on my menus. I offer a charcuterie platter, and it's my best-selling item—my customers won't let me take it off the menu. This proves to me that the work that goes into it is valuable to people. It makes them happy, which is what being a chef is all about to me.

"When I was asked to teach charcuterie at Schoolcraft, my alma mater, I'd been practicing it for nearly twenty years, but when a student asked me, "What's a meat emulsion, how does it work?" I couldn't explain it. It's one of the fundamental charcuterie techniques used in many pâtés, mousselines, and sausages, but I could not put it into words; I didn't really know myself that a meat emulsion, like a mayonnaise, is the suspension of fat in another medium, in this case in protein, with the help of a little water. That's when I dove into the subject, began to study the science and chemistry of it. Here is another level of interest for me—the complex manipulations of fats, proteins, salt, acids, seasonings required for great charcuterie. It's a fascinating science as well as a craft.

"I travel in Europe at least once every couple years, and one of the things that only recently dawned on me is that everywhere I go, charcuterie is part of the local culinary scene. I eat charcuterie everywhere in Europe, it's part of the culture in a way that it isn't here. And the best charcuterie I've eaten so far has been in Italy, the charcuterie of the *salumeria*, the place where dry and semi-dried sausages and dried meats are sold.

"The last time I went to Italy, I took the whole family, and Mom. Eight of us (*otto*). Hot late afternoon, too long a day, everyone cranky, and we're walking around a little town on Lake Como. My wife, Julia, can't find the boys, ages four and six, who are running around somewhere; Alex and Carmen are arguing; my oldest daughter, Alana, is complaining. 'Julia,' I said, 'find the boys, let's just get something to eat.' Very thin nerves all of us. I ducked into the first place I saw, a tiny place, and said to the *proprietario*, "*Otto?*" He immediately shoved two of the five tables together, Julia arrived with the boys, and I noticed that a kid behind the counter had begun slicing. We were still edgy, the kids jostling for the seats they wanted—big headache—and then even before we all sat down, the owner put down a plate with a heap of marinated olives in the center and translucent slices of salami, cubes of mortadella and cured ham, and, on the side, some good bread. It was halting. The way the sunlight hit the fat of the dried meats, the way it glistened, the beauty of the meat. The table was silent. We ate everything. The kids ate everything. All the tension evaporated, gone.

"That moment clinched it for me. Clinched everything I knew about charcuterie and everything I didn't know but need to. It was love, baby! Their love, our love, my love. They welcomed us with a plate of charcuterie, and it made all the difference."

This book explores those techniques of preservation and economy that are scarcely used in today's home kitchens but that result in food that nourishes on many levels. The food is not only delicious to eat, it's also satisfying on an intellectual level. Understanding the culinary mechanisms that cause these great transformations–a plain piece of pork belly becomes bacon or pancetta, ground pork and salt becomes *saucisson sec*, pork shoulder and liver become a country pâté–is a reward in itself. And it's satisfying on a craft level as well, mastering the physical techniques to achieve these transformations.

This food takes some care, some thought, and some common sense. One of the reasons charcuterie techniques are seldom used in either the home kitchen or the average restaurant kitchen is that they're a lot of work. It takes many steps over several days to prepare a confit, for instance. Why not just throw the duck in the oven and roast it? Two or three hours, presto, delicious duck. Crispy skin, succulent meat, maybe a nice sauce. Hard to beat that, so why work so hard?

Because all those steps required for a confit, during each day, are a good thing, are enriching. It's satisfying work if you do it right. To coat the duck pieces with the dry rub is visually appealing; you begin tasting the final dish in your mind. When you wash the dry rub off the duck pieces the next day, you squeeze the flesh and feel how dense it has become overnight from the action of the salt. As the cooking begins, the smell fills the kitchen, and the sight of the duck submerged in the cooking fat is beautiful. As the duck cools, the fat, no longer liquid, gradually obscures the duck. The cooled duck fat is especially unctuous and creamy, a pale deep ivory color. Then, after the duck has chilled in the fridge for a while, and it's a pleasure to see it there whenever you open the door, you remove it from the fat and pop it in the oven. The skin will become very, very crisp, and the meat will be rich and deeply flavored.

Part of doing a recipe like that well means choosing the right time to do it so that you can take your time. Don't try to squeeze the work in between errands. Charcuterie preparations seem easier when you spread them out. If you try to make a pâté twenty minutes before you have to pick up the kids or be at an appointment, you'll be frustrated, and the pâté won't taste good either. Instead, choose a time that you can draw out, extend. Set up your *mise en place*—line up all the ingredients, measured out, and all the tools you need—on a clean counter. The term *mise en place* means put in place, or everything in its place, and refers to all the ingredients and tools you need to get the job done, whether that job is a single recipe or a night's service on the sauté station. Good *mise en place* implies a state of preparedness and organization. Organization is critical to success in the cooking described in this book. It's a good idea to measure out all your ingredients ahead of time, as much as a day before.

Take time to look at the bowl of diced pork that will form the basis of the terrine; the duck breast or pork loin, nicely trimmed, that will be placed inside; and the fresh mushrooms, the minced shallots, the chopped parsley, peeled garlic cloves, the ramekin of kosher salt. It's a tantalizing sight because you know all those ingredients will be transformed in a delicious and dynamic way, and it's going to feed people, your friends and family, in a way that they're not accustomed to. That's part of the pleasure of cooking.

The ingredients you're about to use should look good as they are, individual and organized, before you even get going. If you start sloppy, how can you end clean? Stay organized as you go, and keep your work surface bright.

Enjoy the tools of the craft. For many of these recipes, you'll need some kind of meat grinder. If you do a lot of grinding we recommend you use a meat grinder. But many of these recipes were tested using a KitchenAid mixer with a grinder attachment and a 5- or 6-quart bowl. Could you use a hand-crank grinder? Certainly. Or could you get by with just a food processor? Sure, though the texture of some of the products will be different. Could you make the recipe by hand, with just a knife and a cutting board? For the most part, yes, but it would be more difficult. Having the right tools makes a difference here as it does in any craft.

The individual techniques involved in charcuterie are not hard—you don't have to flute mushrooms, for example, or haul twenty-gallon stockpots around the kitchen—but much of this type of cooking takes plenty of time and good attention. There's a reason these kinds of foods are less common in restaurants than, say, grilled steak, sautéed chicken, or other heat-and-serve items. Life at home, life in a restaurant, is busy. Who has the time? Often we don't, but when we do, when we make the time, there's no more satisfying kind of cooking than this and the highly crafted, intensely flavored food that results.

The art and craft of charcuterie today includes many kinds of what are classified as *garde-manger* techniques (i.e., cold food preparation), such as those for making pâtés and terrines, whether classical (meat or fish blended with fat) or modern (vegetable terrines such as a roasted portabello and red pepper terrine bound with a fresh herb vinaigrette), such as confits and rillettes of duck, goose, pork; and all manner of sauces for countless canapés, appetizers, and first courses.

The backbone of charcuterie is meat or fish that is ground or pureed with seasonings and then cooked, a preparation called a forcemeat, or *farce*, from the French *farcir*, meaning to stuff. But the recipes here encompass a range of techniques that extend beyond classical charcuterie, fundamentals important to many areas of cooking, such as poaching, steaming, sautéing, braising, sauce making, and seasoning. These recipes focus on various uses for proteins, whether egg white, gelatin, meat, or fish; cooking with plenty of fat or cooking very lean; and knowing how and when to do each. Learning the craft of charcuterie extends a cook's abilities beyond those required for charcuterie alone, and it will open new vistas for most home cooks.

This book begins with the importance of, and effects of, salt and with recipes that rely on salt as the primary preserving mechanism. We move on from there to the use of smoke, with recipes that combine salt and smoke (most smoked items must also be cured).

Having paired salt with smoke, we can move on to the sausage—namely, a combination of ingredients that are ground and then simply cooked or smoked in varying ways, with intriguing results. The sausage section, the fat midriff of the book, ends with the complex dried sausages in the style of the Italian *salumeria*. These—salami, soppressata, and the like—rely on a special kind of cooking, one that requires genuine craftsmanship. It's one that reaches a kind of artisanal excellence that's distinct from all other kinds of cooking. It's a culinary specialty that in this country is in its infancy, an art that's slowly emerging in some American restaurants.

From sausages, we move on to the pâté, which is extraordinary for its visual appeal and taste. A pâté or terrine is cool to make, to look at, to serve, and to eat. Yet pâtés and terrines, whether meat, fish, or vegetable, are one of the types of cooking that almost no one does at home anymore.

The last category here is another subspecialty of charcuterie, the cherished confit. And we conclude with a whole chapter on condiments and sauces to be paired with all of these items.

The recipes in this book, with a handful of exceptions, reflect Brian's work as a chef and a teacher. While some are wholly his own, most have their roots in standard preparations that he has molded over the years to satisfy his own tastes and spirit. As we worked on this project, Brian would send the recipes to me in chef-speak. I rewrote them in the style we use here and tested many of them myself. If we didn't like something (this condiment tastes too eggy, do we really need diced sun-dried tomatoes packed in oil here?), we'd discuss the recipe and retest it. And more recently, Brian has used the recipes as they are in this book in his class, further testing and refining them.

The recipes are written in the standard weights and volumes used in the United States, and we've also included metric equivalents. An ounce equals about 28 grams, but for the sake of consistency, we've hued to commonly accepted practical equivalents:

¼ ounce = 7 grams	1 teaspoon = 5 milliliters
½ ounce = 15 grams	1 tablespoon = 15 milliliters
¾ ounce = 20 grams	¼ cup = 60 milliliters
1 ounce = 25 grams	½ cup = 125 milliliters
1 pound = 450 grams	1 cup = 250 milliliters
5 pounds = 2.25 kilograms	1 quart = 1 liter

The recipes range from basic to complex, but the two main charcuterie principles—meat emulsification and salt-curing, as well as what to do with the meat once it's cured (cook it right away, hang it to dry, smoke it, preserve it in fat)—are easy to understand. And all else follows from these ideas.

We want to reiterate that while the techniques used here do not represent quick or casual cooking, they are not terribly complicated. We did not write this book for the restaurant cook. We believe that the techniques of charcuterie can be important and satisfying in any kitchen; and the fact that they are not well known or much used in home kitchens today is unfortunate. We conceived the book primarily for the home cook and the recipes (with the exception of the tricky, dry-cured sausages) are written for the home cook.

Using this Book: Notes about Tools and Ingredients

We recommend the following basic equipment specifically for this kind of cooking, but, except for the sausage stuffer, these are all-purpose players, useful to any cook.

- **A 5- or 6-quart KitchenAid mixer.** Invaluable in the kitchen generally, it is used here extensively for grinding and mixing meat.
- A **meat grinder** is essential for good sausage making. If sausage making or even grinding your own beef for hamburger is a regular part of your cooking year, or if you grind large batches of meat, as many hunters do, we recommend buying a separate meat grinder (see Sources, page 303) rather than a grinder attachment for your stand mixer.
- **A standard food processor** (11-cup/2.75-liter capacity or more).
- **A digital scale,** one of the most valuable kitchen tools any cook can own. Measuring ingredients by weight is indisputably the most accurate form of measuring. Diamond Crystal salt, for example, is flakier and lighter than Morton's kosher salt, which is lighter than Baleine's sea salt—so an ounce of each, by weight, will vary in volume. An ounce of each will have the same effect on the food, but the same volume measure of each one will not. In baking and pastry, weighing rather than measuring flour guarantees far more consistent results. Furthermore, measuring cups and spoons aren't always accurate. So we cannot recommend strongly enough measuring dry ingredients by weight rather than by volume.

  Digital kitchen scales are widely available and can be found starting at about $30. Having a scale that weighs up to five pounds is useful but not critical for these recipes where it's the ounce measures that are most important. It's also helpful to have a scale that converts ounces to grams and vice versa.
- **An instant-read thermometer.** Cooking the food to the right internal temperature, and not beyond, is critical to the success of any dish (not least of which are sausages, which people tend to overcook). Digital thermometers are preferable because of their speed, but analogue instant-reads, which are a few dollars cheaper, are just as good. Digital thermometer-timers are a great convenience. These have a probe attached by a cable to the timer and will sound an alarm when the internal temperature of what you're cooking has been reached.
- **A sausage stuffer.** There are various types of stuffers, but the best are cylindrical ones that have a hand crank. These make the work very easy, quick, and clean, but they're expensive, upwards of $130. KitchenAid makes an inexpensive stuffer that attaches to the grinder; it is acceptable for basic, and occasional, sausage making, but it can be problematic (grinding directly into the casing skips the important mixing step—but some chefs swear by stuffer attachments).
- **A terrine mold.** Le Creuset makes excellent, handsome enameled cast-iron molds with lids, from about $160, but numerous sources offer terrine molds of varying materials that are perfectly fine.

2.

SALT:

HOW THE MOST POWERFUL TOOL IN YOUR KITCHEN
TRANSFORMS THE HUMBLE INTO THE SUBLIME

. . . . .

Salt, whether dry or in a brine, is a powerful rock that not only keeps our bodies alive but has helped us to preserve food for millennia, shaping civilizations as it graces our kitchen table. The following recipes describe how and why it works and some of the amazing food you can create when you use salt as a lever.

Salt, one of those items in everyone's daily life, is completely quiet, resting inside a shaker on the kitchen counter or box in the cupboard, never calling attention to itself, plain, prosaic, no dramatics, humble—a rock, after all. In truth, though, it is a miracle. Its place in the history of civilization attests to this, as does in equal measure its place in the human body. Without the mineral sodium chloride, our muscles would cease to function, our organs would starve. Because our body doesn't produce it and because we need it to survive, humans developed a distinct sense for salt, and our bodies are highly attuned to the need for salt.

Historically, human beings have gotten plenty of salt from meat, and non-meat eating animals have relied on salt licks, or natural salt deposits. Beyond our bodies and beginning with the earliest known civilization, salt, a very concentrated pure form of these two essential elements, sodium and chloride, has been the foundation of entire economies, has provided the means (via food preservation) for a surplus of food on which communities survived, has been glorified in religions, has been a prized catalyst in voodoo and witchcraft, has been used as gift, as symbol, as money. And, Mark Kurlansky notes in his book *Salt*, it's the only family of rocks that humans eat.

Fundamental to salt's rich history is what happens to food when it comes into contact with salt. Most people know that salt enhances the flavor of fresh food, but fewer think about the more important fact that salt preserves food.

It's not time that spoils meat or causes vegetables to rot, it is the bacteria and other microbes feeding on food that do, so if we salt meat or fish, if we soak vegetables in a salt solution, we can disable or kill those microbes and thus dramatically delay, or halt, their destruction of our food. Happily, and perhaps logically, this salt action not only allows us to preserve food, it can make food taste better. A baby cucumber tastes OK, I suppose, but turn it into a pickle, and it's delicious. It goes beautifully with, say, corned beef that has likewise been brined, soaked in heavily salted and spiced water. We originally brined beef to preserve it, but it's so tasty we continue to preserve it just for the flavor. Change that corned beef sandwich to a classic Reuben, with cheese and sauerkraut, and you add two more preserved foods to a single familiar lunch: salt and acids are added to dairy solids to help preserve milk as cheese, and sauerkraut is, of course, salted cabbage.

Cover a side of salmon in salt (and some sugar to counterbalance the harsh effects of the salt) and then a day later, wipe the salt off, slice it paper-thin, and eat it just like that–and it's beautiful. I salt legs of duck for three days if I intend to preserve them in a confit in October to use in a February cassoulet. Salt a raw hog's leg for a couple of weeks, then hang that leg to dry in a cool, humid cellar for months, and you are using the same technique responsible for what are arguably the world's best hams, those uncooked marvels of Bayonne and Parma and San Daniele.

"Most methods are of great antiquity," writes Filipe Fernandez-Armesto of preservation in his food history *Near a Thousand Tables*.

> . . . [F]reeze drying, which most people think of as one of the most up-to-date techniques, was perfected as a way of preserving potatoes by early Andean civilizations over two thousand years ago. The technique was elaborate: overnight freezing, then trampling to squeeze out residual moisture, then sun drying repeated over several days. The durability of frozen food has been known to all Arctic peoples from time immemorial. Wind drying . . . was probably an older technique of preparation than cooking. In every documented period of the history of food, salting, fermenting, and smoking appear among recorded preserving techniques.

The Egyptians were possibly the first people to preserve food with salt on a large scale–they used it not only for their own food supply, but, more important, for trade; it helped to build their economy. (They also used salt to preserve their dead, turning cadavers into mummies.) The Egyptians reviled the pig, and so the invention of the ham was left to the Celts, who thrived in what is now Europe during the Iron Age (around 1000 B.C.). The Celts embraced the pig. They gave it to the Romans, who marched into France and England. The Romans also embraced the ham, and among their favorites was ham cured in Westphalia (now part of Germany) and smoked over local beech and juniper branches. Westphalian ham endures to this day as one of the world's cherished hams.

The Egyptians may have been the first to take the hard, bitter fruit of an olive tree and soak it in saltwater to make that fruit not only edible, but delicious. Again, salt was the key. Salt was plentiful around the Mediterranean, and so salting meat, fish, and vegetables was common from the earliest civilizations in that region.

At the end of the first millennium, the Vikings flourished because they learned how to preserve cod, which sustained them on their journeys. But because salt was not plentiful where they lived, they preserved the fish by hanging it in the chill arid winds of the northern latitudes until it developed the strength of plywood. The Basques, though, at that time, were using salt to preserve their cod; salted cod lasted longer than air-dried cod (and it didn't spoil as did their salted whale meat, which was high in fat, and thus easily became rancid). The Vikings also secretly fished for cod in the New World (keeping the discovery of Nova Scotia to themselves, centuries before Columbus crossed the Atlantic), then salting it to sell throughout Catholic Europe, no doubt a thriving trade on meatless Fridays and during Lent, when pork was off limits. The longer a culture could preserve food, the farther its members could journey and explore; salt pork was common fare on extended voyages. The age of great exploration several centuries later could only have happened after cultures learned how to preserve large amounts of food for long journeys.

Salt's purpose in our body is to help to regulate fluid exchange at the cellular level. It works this way: heightened salt concentration outside a cell (more specifically, electrically charged sodium ions, atomic particles) results in a fluid exchange out of the cell, water moving across the selectively permeable membrane, to reduce the sodium concentration outside the cell and raise the potassium concentration within the cell. Our cells are fed and nourished through this fluid exchange.

The key here is the selectively permeable membrane. Such a membrane allows certain kinds of molecules to flow through it, such as water and salt and other electrolytes, but not bigger molecules, such as proteins. A semipermeable membrane is like a tea bag, which lets water pass through it, but not the tea leaves.

When we put a piece of meat into an environment in which the salt concentration is very high (as with a brine or dry cure), the same exchange happens in the cells of the meat as happens in our body: Attracted by sodium's ions, water rushes out of the cells to join them. Equilibrium is always sought, so there is a continual back-and-forth movement across the membrane as the concentrations shift and salt in solution enters the cells of the meat (bringing some flavoring) and returns to the brine or cure (along with blood). The ionic charges also change the shape of the proteins, loosening them and allowing them to contain more moisture.

By pulling water out of the meat, salt, by definition, dehydrates it. When it enters the cells of the meat, it also dehydrates the microbes that cause decay and spoilage and other potentially hazardous bacteria, either killing them or inhibiting their ability to multiply. This is salt's main preservative mechanism—dehydrating microbes. A secondary preservative effect is that it reduces the amount of water in the meat, which microbes need in order to thrive.

Virtually every food group can be salted to excellent effect. Eggs can be pickled (a pickle is a brine that includes a strong acidic component). Most fruits and vegetables can be preserved through some sort of salting, sugaring, or pickling.

Of all the world's foods that can be preserved to great effect, the pig has proved to be by far the most versatile. It is the only animal that has generated its own culinary specialty: charcuterie.

In the same way that salt is a kind of unsung marvel, the pig is an animal whose glories go largely unrecognized in America. In France they like to say that every part of the pig is used except the oink. That's not quite true. The utilitarian French scald off the bristles, and they have yet to find a use for the toenails. But other than that, everything is used. Furthermore, the pig provides a range of widely differing things to eat, more in fact than any single other animal we know of. Compare it to beef, for example, where you've got everything from the tenderloin to the tough shank–but it's all pretty similar in taste, whether filet mignon, stewing beef, or hamburger.

The pig, on the other hand, gives us ham, fresh sausage, tenderloin, chops, ribs, hocks, trotters, and blood for boudin noir, all of them with distinct differences in flavor and texture. Its liver is superior in pâtés. Its belly is fantastic, especially salted and smoked–pork takes the flavor of smoke like nothing else–the preparation we call bacon. The belly is best of all, in my opinion, when it's salted, then confited. Pig's feet are loaded with gelatin, which enriches stocks and produces aspic, and the meat from the feet and shanks is otherworldly when braised and seasoned. Even the skin, if cooked long and slow, becomes succulent and delicious and, diced, is an excellent addition to many dishes.

The rendered fat of the pig is soft and pure and creamy; indeed, for centuries, that rendered pig fat, which we call lard, was one of the pig's most valued attributes. It can be used as a cooking fat or as a shortening in pastries both savory and sweet. Replace the vegetable shortening in a traditional pie dough with half butter and half good lard, and the result will be a beautiful golden brown crust that's flavorful, crisp, and flaky. (I can find lard, freshly rendered by the Amish farmers who grow hogs, during the summer at a growers' market, but I can also purchase it by mail or order fat from my meat department and gently render it myself–see page 262 for the technique). Replacing lard with hydrogenated vegetable oil for the shortening in cookies and crusts has resulted in pastries that are a shadow of their former selves.

Still another advantage of pork is that it takes to just about any form of cooking. Grilled or roasted, sautéed or braised, pork is fantastic.

And salt and pork are a love affair, a marriage resulting in still more flavors and textures, depending on what part of the animal is salted, whether belly or ham, or even the fat itself. The flavor of pig fat is neutral: it's soft but firm, creamy in texture (unlike beef fat, suet, which is hard, an indication of its higher saturated fat content). While at a restaurant in the mountains of Carrara, Italy, in 1988, my wife and I were asked if we'd like to try the house specialty, called *lardo*. The proprietor, Fausto, who had been feeding the workers from the legendary marble quarries for years, salted and seasoned thick slabs of pork back fat and cured them for months in marble casks. He served us three thin slices of this cured raw fat on some basil leaves, with olive oil, salt, and pepper. Soft, creamy, and delicious, it was a revelation.

Lardo has recently been making its way into American restaurants, though usually disguised under names such as *carpaccio bianco* and other innocent-sounding aliases (but pure pig fat by any other name is just as delicious). Pig fat is various too: Fat from the jowl, like the thick back fat, is creamy and excellent for pâtés and sausages. Some cooks prize leaf fat, which surrounds the kidneys, most highly of all for use as lard.

I think of Fausto's *lardo* longingly, and not without irony and sadness, living in a country where so many people claim to avoid fat and salt as if they were evil incarnate and yet think nothing of devouring sodium-rich, fat-laden fast foods that come in boxes and bags.

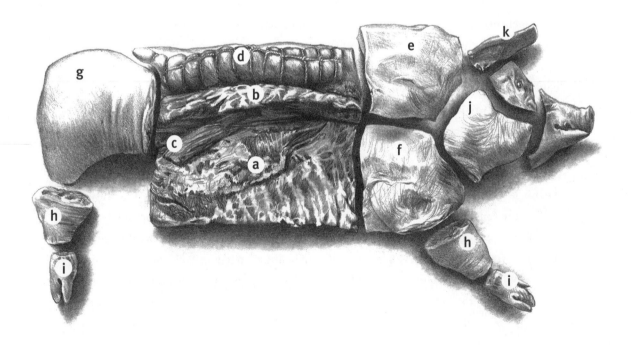

Interestingly, the way a pig grows naturally, its ratio of fat to meat is culinary perfection. What is considered to be the optimal ratio of fat to meat in sausage is 30 percent to 70 percent, which is pretty much the composition of a pork shoulder.

There are recipes and descriptions of all varieties of food in this book, for vegetables and fish, and all manner of sauces and condiments, but we'd like to make sure that one thing is understood here and now: The pig is king.

KEY TO PARTS OF THE PIG

This detailed drawing (*opposite*) allows you to see where the bacon, the long flat slab of belly, is (a). Notice the striations of fat and how they change as they move from shoulder to ham; if you buy a belly with ribs still attached, these striations are where the ribs will be connected; the back ribs, or baby back ribs (b), connect to the spine. Extending from the spareribs will be a wide strip of striated muscle, the diaphragm, or the skirt steak (the ribs and diaphragm have been removed in this drawing). The skirt steak from a hog can be braised or slow-roasted, like spareribs, and is wonderfully flavorful, succulent, and tender, or it can be sliced thinly and stir-fried.

Drifting down from the ribs toward the ham is the pork tenderloin (c), the muscle nestled in the underside of the back ribs. Because it is not strenuously used, it is very tender and so needs little cooking.

The large long muscle riding on top of the ribs is the pork loin (d). Brian has tied it to be roasted. Notice how low the ribs are and how tall the spine. Normally the loin is covered by a very thick layer of back fat. You can get a sense of the thickness of the back fat in the shoulder piece to the right of the loin.

The shoulder butt (e) is the muscular fatty cut from which sausage is made. Below it is what is called the picnic ham (f). Both the shoulder and picnic ham must be cooked low and slow before they will become tender. When working with these cuts, it's important to be on the lookout for glands scattered between the shoulder and the head. About the size of a quarter, they're more brown than pink and squishy rather than firm, neither fat nor flesh. These should be trimmed away and discarded.

At the back end of the hog is the ham (g). This is the cut that can become the great *jamón ibérico* or prosciutto or Smithfield ham (or even a honey-baked ham). Each pig, of course, has two of them. Between them, not pictured here, is the tail, which can also be braised to good effect.

Extending from each picnic ham and ham is a shank, or hock (h). Ham hocks are best known smoked, used to flavor stews and the like, but they can be very good simply braised. Extending from the hocks are the feet, the trotters (i). These pieces are filled with a lot of bones and very little meat, but they can be cleaned out, stuffed and braised for the eponymous bistro dish. (This dish is actually much better made from hock meat.)

The jowl (j) is a wide slender slab of meat and fat that can be cured like pancetta and hung to dry, then used like pancetta or bacon or sliced paper-thin and eaten as is. It's fantastic, and probably the easiest meat to cure for the home cook. It's especially important with the jowl to check the meat for glands.

The ears (k) can be braised, peeled, and julienned, then used as garnish in pâtés and headcheese, which is made by simmering the entire head until the meat is falling off the bone, then chilling the shredded meat and julienned ears with vegetables and herbs with some of the highly gelatinized stock in a terrine mold.

LIVE WEIGHT: BUYING A WHOLE HOG

At a local growers' market on the eastern edge of Cleveland, I got to know an Amish family selling a variety of grains, eggs, fruits, vegetables, and pork. I asked Daniel, the one responsible for raising the hogs, if we could buy one whole. I wanted to work with some of the more difficult-to-obtain items, such as intestines and blood. I also wanted to see the difference firsthand between the factory-raised hogs and a farm-raised hog, one that had fed on the apples and nuts and corn that grew at Daniel's farm.

Daniel said he'd charge me eighty cents per pound, live weight. I spoke with Brian about this, and he said he'd like one too. By late October, our hogs weighed just under 300 pounds. (For the record, pigs can be called hogs when they reach sexual maturity at about seven months or weigh over 160 pounds.) With a small fee for delivering them to the butchery, and a fee by the butcher to take the bristles off the skin, stun, kill, clean, and halve the hogs, the total cost for each of us came to about $275.

I was able, just, to fit the two slaughtered beasts in the back of our Jeep Grand Cherokee, covered with sacks of ice, and I took them to our friends Mark and Giovanna Daverio, who own Battuto, our favorite Italian restaurant. Giovanna used to work in the kitchen of Zuni Café, cooking with the legendary Judy Rodgers, and Mark had worked at Oliveto under Paul Bertolli, whose skill with hogs is probably unparalleled among American chefs. We enlisted their help, or rather their walk-in cooler, to store the hogs before Brian arrived to break them down. It was very much a *Sopranos* moment, hefting the beasts into a restaurant that happens to be in the center of Cleveland's Little Italy.

Brian arrived the following day to demonstrate for us butchering the whole hogs, something he does every couple weeks for his students. In addition to the education of working with steaming innards, which is exactly as disgusting as you would imagine, here's what we had when we broke each hog down into its various cuts.

2 jowls for drying, about 1 pound/
450 grams each
2 picnic hams, 8½ pounds/3.86 kilograms each
2 shoulder butts, 11 pounds/5 kilograms each
2 hams, 23 pounds/10.5 kilograms each
2 bellies, 14 pounds/6.3 kilograms each

**Two 30-pound/13.6-kilogram loin sections, which each included
an 8-pound/3.6-kilogram loin roast, 2 tenderloins, and
a good 10 pounds/4.5 kilograms of back fat
16 pounds/7.25 kilograms trimmings to grind for sausage
15 pounds/6.8 kilograms fat
10 pounds/4.5 kilograms skin
32 pounds/14.5 kilograms bones for stock
about 1 gallon/4 liters blood**

In all, a little more than 200 pounds/93 kilograms of usable product per hog, or $1.35 per pound.

Beyond the good value and a better understanding of the whole animal, though, was seeing the difference in quality and size between this naturally grown hog, with its dark pink, well-marbled muscle, deep pork flavor, and copious smooth, supple fat, and commercial pork. The belly was a good three inches thick; none of us had seen one so fine.

Another great pleasure was making a proper blood sausage (see the recipe on page 143). We used the fresh hog intestines for this (cleaned as described on page 103), which we felt was important. We diced and cooked apple and onion and seasoned it, blended it all in a large steel bowl, and added the blood. We funneled the pudding-like mixture into the hog's intestine, marveling at the extraordinary lavender hue, then cooked it and ate it later that evening. It was mild and sweet from the onion and apple, the blood and egg acting as a binder, like the aspic in headcheese, and the blood a flavoring ingredient rather than a main item. But if you take a poetical view of the world, the triumph of the sausage was to fill the hog's intestine with its own blood and with the food that was part of its natural diet (it had feasted on apples and onions and it continued to do so, in a manner of speaking). If you don't take a poetical view, then this must do: It was delicious.

We ate the finer fresh cuts—tenderloin and loin—fairly quickly, but the garlic sausage we made lasted for many weeks, ready to be grilled or roasted or sliced into a white bean and escarole soup that fed scores of people at a Christmas party. I made bacon with part of the belly and confited the rest, slow-slow-roasted the shoulder for pulled pork. The hams (along with many other cuts) went to Mark and Giovanna, who cured them and hung them to dry for prosciutto.

For any home cook who has access to locally raised hogs, and a second refrigerator and freezer for storage, this is a great and exciting option.

The following recipes feature salt's power to transform food into preparations of extraordinary flavor and texture. Salt appears in two forms in the recipes–dry, in which the salt is simply rubbed onto meat or fish, and wet, in which the salt is dissolved in water. All of the recipes are for foods once valued for their capacity to be preserved by salt, but which we now cherish because they're so good to eat–things like bacon, cured salmon, and corned beef.

Dry Cures

Curing means to preserve meat or fish with salt. For dry-curing as opposed to curing in a brine, a salt mixture is simply rubbed over the meat or the meat is dredged in it.

An important note about salt measurements: It's best to weigh salt rather than to measure by volume because salts differ in weight by volume. I use Morton's Kosher Salt; a cup weighs almost 8 ounces. Brian uses Diamond Crystal kosher salt; a cup of this salt weighs 4.8 ounces. That's a big difference. If you do not have a scale to weigh your salt, we recommend using Morton's Kosher Salt for these recipes.

There are different kinds of salts, of course, but throughout this book salt almost always refers to sodium chloride. What we call curing salts have nitrite in them and sometimes nitrate as well. Nitrite does a few special things to meat: it changes the flavor, preserves the meat's red color, prevents fat from developing rancid flavors, and prevents many bacteria from growing, most notably those responsible for botulism poisoning. Curing salt with nitrite is called by different names and sold under various brand names (tinted cure mix, or T.C.M., Insta Cure #1, and, our preference, DQ Curing Salt). We call it pink salt because that's what it looks like and how it's commonly referred to in restaurant kitchens. No matter the name, it's all the same: 93.75 percent salt and 6.25 percent nitrite. Nitrites, which are found in green leafy vegetables such as spinach and root vegetables, are not harmful or dangerous in small quantities, but in large quantities they are, and the curing mixture is dyed pink to prevent its accidental use or consumption; please treat it appropriately. Its most important function is to prevent botulism poisoning from sausages and other foods that are smoked. Sometimes nitrate is added to pink salt; this curing salt is used for dry-cured sausages, sausages that cure for a long time, and is sold under the brand names DQ Curing Salt #2 and Insta Cure #2. Saltpeter (potassium nitrate) was traditionally used to cure meat (and still is in Europe), but its effects are less consistent than today's commercially manufactured curing salts. For further discussion of nitrite and nitrate issues, see pages 174–176.

When curing meat to preserve it, the time the meat or fish spends in the cure is critical.

Meat and fish that sit too long in a dry cure can become too salty. But the bigger the piece of meat, the longer it needs to stay in the cure, so that the cure can penetrate to its center.

A general working ratio for a dry cure is 2 parts salt to 1 part sugar, plus 10 percent of their combined weight of pink salt (1 ounce/25 grams of pink salt is enough for 25 pounds/11.25 kilograms of meat). Salt is the critical active ingredient; sugar is important in that it compensates for the harshness of the salt. You can add various seasonings to the cure or alter its sweetness, depending on your taste and what you're curing. Bacon, for example, benefits from sweetness, so you might add brown sugar or maple syrup. If you prefer a more savory bacon, add garlic and black pepper. If you don't want sweetness—say, for a turkey or chicken or pork loin—then you can reduce the amount of sugar or add savory aromatics.

THE BASIC DRY CURE

The following basic dry cure can be used to make any kind of cured product, but it is especially fine with pork. You can use either what's called the "salt box method," which doesn't require measuring, or measure out 2 ounces/56 grams of this mixture for each 5 pounds/2.25 kilograms meat. The salt box method means simply dredging the meat in plenty of dry cure on all sides, then gently shaking off the excess so that it has an even coating of dry cure.

Brian and I prefer to use dextrose, a refined corn sugar, rather than table sugar because it is less sweet and, because as the grains are very fine, it dissolves more easily and therefore has a more uniform distribution. But granulated sugar is fine as well.

THE BASIC DRY CURE WITH GRANULATED SUGAR
1 pound/450 grams kosher salt
8 ounces/225 grams sugar
8 teaspoons/56 grams pink salt

THE BASIC DRY CURE WITH DEXTROSE
1 pound/450 grams kosher salt
13 ounces/425 grams dextrose
8 teaspoons/56 grams pink salt

Combine all the ingredients, mixing well. Stored in a plastic container, this keeps indefinitely.

Yield: About 3½ cups/725 grams if made with granulated
sugar, 4½ cups/950 grams if made with dextrose

Simple Bacon at Home

Bacon, which comes from the belly of the hog, is good fried in strips and served with eggs, but, more broadly speaking, it's one of the great flavors available to a cook, and depending on how the bacon itself is prepared, it can be a versatile ingredient in many dishes. Because in this country, even at meat counters, bacon is almost always sold in strips, we are less likely to take advantage of its versatility. When you make your own, however, you can cut strips ½ inch/ 1 centimeter wide and then cut those into the perfect batons called *lardons*, ½ inch/1 centimeter by ½ inch/1 centimeter by 1½ inches/3.5 centimeters—the size you need in order to cook them so they're crispy on the outside with a chewy interior. Then you can make a real coq au vin or a traditional frisée salad with a poached egg and lardons. Bigger chunks can be cut for stews, soups, and bean dishes. Slabs can be brushed with mustard and honey and roasted whole. Fresh bacon gently grilled is an extraordinary treat.

Happily, bacon is very easy to make at home. In this country, bacon is by definition smoked after it's cured, but the smoke is really a secondary flavor, like a seasoning. The genuine bacon flavor comes partly from the sodium nitrite in the cure. If you have a smoker, or if you can create some low-heat smoke in a kettle barbecue, that will deepen the traditional bacon flavor, but no special equipment is necessary to cure your own bacon at home.

Furthermore, what you make at home will be superior to just about anything you can buy at supermarkets. Most of the bacon there comes from factory-raised hogs, the curing done at commercial plants, and the result is thin strips of watery meat that, even when cooked until crisp, have a taste only reminiscent of real bacon.

When you make your own bacon and fry a slice, you'll know what bacon is all about. Notice the copious amount of fat that renders out, and that the meat doesn't reduce in size by fifty percent. The result can give you an understanding of why bacon became such a powerful part of America's culinary culture. The chefs and butchers who cure it as a part of their work may be able to ensure that we don't lose it, but the tradition could be reinvigorated if more home cooks cured their own bacon, and then roasted slabs of it, slicing large chunks to serve as a garnish for a roasted loin, say, or confit it, or cut big lardons for salads. Or just served proper bacon for breakfast.

Slicing home-cured bacon as thin as supermarket bacon can be difficult, but home-cured bacon ought to be dense and chewy, so don't worry too much about thinness (it helps to use a slicing knife—which has a long, thin blade—or to freeze the bacon before slicing). Left-over trimmings are fantastic in stews and sauces. Because of its high fat content, bacon keeps well frozen, making it easy to always have some on hand to throw into the pot or sauté pan.

FRESH BACON

Pork belly and pink salt are the two special items you need to make bacon, everything else is usually on hand. Some specialty markets may sell pork belly regularly—and in that case, you can specify exactly how much you want, say 3 to 5 pounds/1.5 to 2.25 kilograms—but in most parts of the country, you will have to order it through the meat department, which is easy enough to do. Excellent pork belly can also be ordered online. Either way, you'll most likely receive a slab of between 5 and 10 pounds/2.25 and 4.5 kilograms. The pink salt, which is inexpensive and lasts a long time, often must be mail-ordered (see Sources, page 303).

Fresh bacon is the simplest and purest kind of bacon to make, with a very mild flavor. Coat the slab of belly with the Basic Dry Cure, refrigerate it for about seven days or so, depending on its thickness, then rinse and pat dry. That's it—you're good to go. You can slice and cook it as is, sautéing it very slowly, and it's delicious.

Traditionally, however, once it is cured, bacon is hot-smoked to a temperature of 150 degrees F./65 degrees C., then cooled and sliced. Because most people don't own smokers, we suggest roasting the cured bacon in a low oven to that same temperature, which will take about two hours. Alternatively, you can grill the cured belly very slowly over indirect heat for a deeper smoky flavor. In either case, the result will be delicious fresh bacon that can be sliced and fried, cut into lardons, cut into chucks for stews, or roasted or grilled whole. Stored in the refrigerator, it will keep for up to two weeks. Or it can be well wrapped and frozen for up to two months or longer.

Variations on the cure are simple. Add maple syrup or maple sugar for sweeter bacon (see below). Or, if you like a more savory bacon, add smashed garlic cloves, bay leaves, and plenty of cracked black pepper. Go a little further in the savory direction with juniper berries, herbs, and nutmeg, and you'll have pancetta on your hands (see page 44). Generally, if you intend to slice your bacon and cook it in strips, the sweeter cure is better. If you intend to use the bacon for a variety of other preparations (lardons in a coq au vin or frisée salad, chunks in a tomato sauce, julienne in a carbonara), you may want to take your cure in the more savory direction.

A final note: This recipe calls for dredging the belly in the salt cure so that all sides are evenly and well coated, no matter what the shape or weight of the belly is. If your belly weighs between 3 and 5 pounds/1.5 and 2.25 kilograms, it's fine to simplify the method by placing the belly in the Ziploc bag, adding ¼ cup/30 grams of dry cure along with whatever additional sugar and seasonings of your choice, closing the bag and shaking it to distribute the ingredients. It's no more complicated than that.

One 3- to 5-pound/1.5- to 2.25-kilogram slab pork belly, skin on
Basic Dry Cure (page 39) as necessary for dredging (about ¼ cup/50 grams)

OPTIONAL
For sweeter bacon, add ½ cup/125 milliliters maple syrup or ½ cup/125 grams maple sugar or packed dark brown sugar; for more savory bacon, add 5 smashed cloves of garlic, 3 crushed bay leaves, and 1 tablespoon/10 grams black peppercorns, partially cracked with the bottom of a heavy pan or side of a knife

1. Trim the belly so that its edges are neat and square. Spread the dry cure on a baking sheet or in a container large enough to accommodate the belly. Press all sides of the belly into the cure to give it a thick uniform coating over the entire surface.

2. Place the belly in a 2-gallon/8-liter Ziploc bag or a covered nonreactive container just large enough to hold it. The pork will release a lot of liquid as it cures, and it's important that the meat and the container are a good fit so that the cure remains in contact with the meat. The salty cure liquid that will be released, water leached from the pork by the salt, must be allowed to surround the meat for continuous curing. The plastic bag allows you to redistribute the cure (technically called *overhauling*) without touching the meat, which is cleaner and easier. Refrigerate the belly for 7 days, flipping the bag or meat to redistribute the cure liquid every other day.

3. After 7 days, check the belly for firmness. If it feels firm at its thickest point, it's cured. One week should be enough time to cure the bacon, but if it still feels squishy, refrigerate it for up to 2 more days. (Belly from a factory-raised hog may be thin, only an inch/2.5 centimeters or so; belly from a farm-raised hog, always preferable, may be as thick as 3 inches/7.5 centimeters.) The thicker the belly, the longer it will take to cure.

4. Remove the belly from the cure, rinse it thoroughly, and pat it dry with paper towels; discard the curing liquid. It can rest in the refrigerator, covered, for up to 3 days at this point.

5. Preheat the oven to 200 degrees F./93 degrees C.

6. Put the belly in a roasting pan, preferably on a rack for even cooking, and roast until it reaches an internal temperature of 150 degrees F./65 degrees C., about 2 hours; begin taking its temperature after 1½ hours. It will have an appealing roasted appearance and good aroma, and it will feel firm to the touch. Remove the rind or skin, now, when the fat is still hot, using a large sharp chef's knife.

7. Allow the bacon to cool to room temperature (try a piece now though, straight out of the oven–it's irresistible; remember that end pieces may be a little more salty than the rest). Once it is cool, wrap well and refrigerate.

8. When the bacon has chilled, slice off a small piece, gently cook, and then taste for flavor and seasoning. If the bacon has cured too long and is too salty, it's unfortunate but fixable; blanching the bacon in simmering water for 1 minute before cooking it, will reduce the salt content considerably. Blanched bacon also tends to crisp up especially well, and lardons are best blanched before being sautéed, for the same reason, regardless of salt content. If you will be using the bacon in stews, though, you don't need to blanch; just be cautious when seasoning the stew.

9. Refrigerate again until ready to use. The bacon will keep for 1 to 2 weeks refrigerated. If you don't plan to use it all during that time, cut it into slices, lardons, and/or chunks, wrap it well, label and date it, and freeze for up to 3 months.

Yield: 2½ to 4 pounds/1.25 to 2 kilograms bacon; 12 to 16 servings

PANCETTA

Pancetta is an Italian bacon and a delicious ingredient used in many of that country's dishes. Like the Fresh Bacon (page 41) it's simply pork belly cured with salt and seasonings, which is then rolled into a log and hung to dry for a couple weeks. It's typically thinly sliced or diced and sautéed, then combined with sautéed vegetables. Countless recipes begin with the gentle sautéing of onions and other aromatic vegetables; precede this step by sautéing diced pancetta, and you'll add a layer of great complexity to the dish. The classic Roman dish spaghetti alla carbonara is made with sautéed pancetta and eggs. Chunks of pancetta can be added to stews, beans, and soups. Cabbage and Brussels sprouts are superb when sautéed with pancetta.

Marcella Hazan, in *The Essentials of Classic Italian Cooking*, notes that pancettas "savory-sweet unsmoked flavor has no wholly satisfactory substitute." She suggests rolling it up in veal scaloppini, then sautéing the rolls in butter and serving them with a tomato sauce, or sautéing it with spring peas (a traditional preparation), or braising Boston lettuce with it.

The traditional process of curing and drying pancetta takes about three weeks, but variations here are a matter of taste. You can reduce the drying time to two or three days, or eliminate it altogether (the pancetta will still taste delicious when cooked). You could also choose not to roll it and use it as is, treating it as you would fresh bacon.

One 5-pound/2.25-kilogram slab pork belly, skin removed

THE DRY CURE
4 garlic cloves, minced
2 teaspoons/14 grams pink salt
¼ cup/50 grams kosher salt
2 tablespoons/26 grams dark brown sugar
4 tablespoons/40 grams coarsely ground black pepper
**2 tablespoons/10 grams juniper berries, crushed with the
        bottom of a small sauté pan**
4 bay leaves, crumbled
1 teaspoon/4 grams freshly grated nutmeg
4 or 5 sprigs fresh thyme

1. Trim the belly so that its edges are neat and square.

2. Combine the ingredients for the cure in a bowl, reserving half of the black pepper, and mix thoroughly so that the pink salt is evenly distributed. Rub the mixture all over the belly to give it a uniform coating over the entire surface.

3. Place the belly in a 2-gallon/8-liter Ziploc bag or in a covered nonreactive container just large enough to hold it. Refrigerate for 7 days. Without removing the belly from the bag, rub the belly to redistribute the seasonings and flip it over every other day (a process called *overhauling*).

4. After 7 days, check the belly for firmness. If it feels firm at its thickest point, it's cured. If it still feels squishy, refrigerate it on the cure for 1 to 2 more days.

5. Remove the belly from the bag or container, rinse it thoroughly under cold water, and pat it dry. Sprinkle the meat side with the cracked pepper. Starting from a long side, roll up the pork belly tightly, as you would a thick towel, and tie it very tightly with butcher's string at 1- to 2-inch/2.5- to 5-centimeter intervals; it's important that there are no air pockets inside the roll (it can't be too tightly rolled). (Alternately, the pancetta can be left flat, wrapped in cheesecloth, and hung to dry for 5 to 7 days.)

6. Using the string to suspend it, hang the pancetta in a cool, humid place to dry for 2 weeks. The ideal conditions are 50 to 60 degrees F./8 to 15 degrees C. with 60 percent humidity, but a cool, humid basement works fine, as will most any place that's out of the sun. (I often hang mine in our kitchen next to the hanging pans on either side of the stove.) Humidity is important: If your pancetta begins to get hard, it's drying out and should be wrapped and refrigerated. The pancetta should be firm but pliable, not hard. Because pancetta isn't meant to be eaten raw, the drying isn't as critical a stage as it is for items such as prosciutto or dry-cured sausages. But drying pancetta enhances its texture, intensifies its flavor, and helps it to last longer.

7. After drying, the pancetta can be wrapped in plastic and refrigerated for 3 weeks, or more, or frozen for up to 4 months. Freezing makes it easier to slice thin.

Yield: 4 pounds/1.75 kilograms pancetta

MAKING PANCETTA

1. The rough edges of the pork belly should be squared off so that it has clean flat edges and a neat square or rectangular shape. The scraps can be used to make Salt Pork (page 47) or frozen for making sausage (fat and meat from the belly are delicious in sausage).

2. The cure should be rubbed liberally over the entire surface of the pork belly.

3. When the belly is completely cured, it is thoroughly rinsed under cold water to remove the cure, salts, sugar, and any aromatics. Some pepper or other seasonings may remain stuck to the meat and fat—this is fine.

4. After the belly has been cured and thoroughly rinsed, the meat side of the belly is seasoned again with coarsely ground pepper.

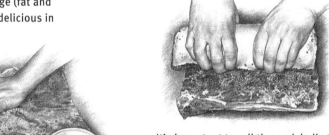

5. It's important to roll the pork belly tightly to avoid trapping any air inside the roll.

6. The pancetta is tied tightly to secure its shape.

7. Cross section of finished pancetta.

SALT PORK

Salt pork is made by curing chunks of pork with a dry cure or in a brine. Making salt pork is the perfect use for the trimmings from a pork belly or pork shoulder.

Salt pork was one of the most important cured items in Europe, especially so in the age of great exploration because, properly handled, it would last in its brine for years at room temperature. When the cook wanted to use it, he simply removed a piece from its brine, soaked it in water, and simmered it long and slow. Today salt pork is rarely used by itself, but it is a key seasoning in traditional clam chowders and a flavoring ingredient in many stews and braises.

You can preserve the pork in just kosher salt and sugar but without pink salt, it won't have the same rosy color or cured flavor. When making bacon or pancetta, it's always good to square off the piece of belly for a neater finished appearance. If you square it off generously, you'll have large pieces rich in fat for salt pork. Salt pork can also be smoked after it is cured. Stored in the freezer, chunks of salt pork are excellent for impromptu use in stews, soups, and bean dishes. Throw a chunk into your tomato sauce for spaghetti, and the sauce will acquire a richness that will surprise you. And it's a must for classic New England clam chowder.

Stored in a freezer bag or tightly sealed container in plenty of salt, salt pork will keep in your freezer for at least a year, but because it may begin to pick up other odors after too much time, I find it's best used within three or four months. And it's true to its name, so when you use it in soups and stews, wait until the salt has dissolved into the cooking liquid before seasoning the dish.

Pork belly or boneless pork shoulder, cut into 2-inch/ 5-centimeter chunks
Basic Dry Cure (page 39) necessary for dredging (about ½ cup/125 grams per pound/450 grams of meat)

1. Dredge the pieces of pork in enough cure to completely coat them (you'll use about 2 tablespoons/25 grams or less per 1 pound/450 grams, but dredging the meat in plenty of dry cure ensures that the pieces will be uniformly coated).

2. Place the pieces in a nonreactive container. Cover and refrigerate for 6 days.

3. Toss the pieces of pork to ensure even distribution of the cure, cover, and refrigerate for 6 more days.

4. Rinse the meat thoroughly and pat dry. Refrigerate in a fresh Ziploc bag or tightly covered container for up to 3 weeks, or freeze for several months.

SALT COD

Cod was one of the most important preserved foods in early civilizations, in part because it could be preserved without salt, as the Vikings, who flourished between the eighth and tenth centuries, discovered. This was a significant discovery, because the Vikings didn't have access to large quantities of salt: air-dried cod provided them with vital nourishment on their long journeys. When the Basques discovered cod (perhaps introduced to them by Vikings), they salted it, and salted dried cod lasted even longer than cod that had simply been air dried.

Salt does the same thing to cod as it does to pork: it reduces its water content and creates an inhospitable environment for the bacteria that could spoil the flesh. Pink salt (sodium nitrite) is not used to salt cod, but the fish is dried after being cured.

Salt cod stacked on tables like lengths of tree bark is a typical sight in Italian, French, Spanish, Portuguese, and Caribbean markets. Commercial salt cod is available in America at specialty stores, but its quality varies. Salt cod must always be soaked for at least a day in a few changes of water to remove the salt and rehydrate the flesh. When it is reconstituted, the texture is denser and the flavor is more concentrated than those of fresh cod. To be blunt, fresh cod has a neutral, even bland flavor; salt cod has great taste.

The reason to salt your own cod rather than to buy it is to enjoy a much fresher, more succulent piece of fish, one that's good enough to serve whole as a main course. Also, depending how it was salted and stored and for how long, commercial salt cod may also have additional unwanted flavors, so salting your own ensures consistent quality. Some fine restaurants, such as Zuni Café in San Francisco and Bouchon in the Napa Valley, routinely salt their own cod.

The following method for salt cod uses plenty of salt to fully coat and dehydrate the flesh. Cod preserved this way will keep for many months refrigerated, wrapped well to protect it from other odors. It must be reconstituted and leached of salt for 24 hours in at least three changes of water; the reconstitution process is complete when the flesh is pliable and just slightly firmer than when fresh.

Once it's reconstituted, the cod can be used in a traditional brandade (mashed with garlic, shallot, olive oil, and some cooked potato, if you wish, and served with slices of crusty toasted baguette), or shaped into cakes and sautéed. Some people like to eat it as is, no additional cooking, with olive oil. Whole pieces of salt cod can be broiled, simmered in tomato sauce, or poached (very gently) in water, milk, or olive oil. Look to Provençal, Spanish, Portuguese, or Catalan cookbooks, or the books from the aforementioned restaurants, for specific recipes, or experiment on your own. It's hard to go wrong with this versatile fish.

One 2-pound/1-kilogram fresh skinless cod fillet
Kosher salt for dredging (about 1 cup/225 grams)
Cheesecloth

1. Dredge the fish generously in salt, pressing it firmly into the salt so that the entire surface is completely and evenly coated.

2. Wrap the cod in two layers of cheesecloth and place on a rack, or in a perforated pan, set over another pan or tray to catch the juices. Refrigerate the fish, uncovered, for 24 hours for every inch of thickness.

3. Remove the cod from the cheesecloth and rinse the salt off thoroughly. Pat it dry, and wrap it in a single layer of fresh cheesecloth. Refrigerate, uncovered, for 4 to 7 days, preferably on a rack to allow air circulation on all sides, until it is completely firm. The time will vary considerably depending on the thickness of the fillet. It will become slightly more opaque than when it was fresh, and it should feel completely stiff through to its center.

4. If you don't intend to use the cod immediately, remove the cheesecloth and wrap the dried fish in plastic. It can be refrigerated for up to 2 or 3 months.

5. When ready to use the fish, soak it for 18 to 24 hours in plenty of fresh water, changing the water every 8 hours or so.

Yield: 1½ pounds/675 grams salt cod

FENNEL-CURED SALMON

Salmon is a commonly and easily cured fish, but what's special about curing salmon is that the process can introduce other flavors. Then, when cured and sliced very thin, the fish has a delicious flavor and appealing texture.

Traditionally, the salmon is packed in a mixture of salt and sugar and cured for one to three days, depending on its thickness. Optional flavors and seasonings are up to the cook.

Here the salmon has the anise flavors of fennel and Pernod. Citrus flavors, added in the form of lemon and orange zest and juice, make for a very bright, fresh-tasting cure. Other common cures include pastrami seasonings (black pepper and coriander); dill, as in traditional gravlax; and horseradish. There are no hard-and-fast ratios for the amounts of such ingredients, but general guidelines would be the zest of two lemons and two oranges and a squeeze of juice from each; an even coating of pepper and coriander; a generous bunch of dill; or a half cup or so of grated fresh horseradish.

As a rule, a 2- to 3-pound/1- to 1.5-kilogram side of salmon requires 6 to 8 ounces/ 170 to 225 grams of salt and between half to an equal weight of that in sugar. This recipe calls for extra sugar, which results in a sweeter, moister cure. This type of cure is known as a soft cure because of its high sugar content. You can cure just a pound of salmon if you wish, reducing the recipe by half. The key element is the length of time in the cure rather than the amount of cure. This cure works perfectly for a piece of salmon that's about 1½ inches/3.5 centimeters thick. If you buy a 1-pound/450-gram tail piece, which will be much thinner, the curing time will be reduced to about a day. (Use touch to determine when it's done, as described below.)

Home-cured salmon is very easy to do and it's a treat for people who love salmon. Be sure to start with very fresh, preferably wild, salmon with brightly colored, firm flesh that smells clean and appealing.

4 ounces/125 grams sugar (about ½ cup)
6 ounces/180 grams light brown sugar (about 1 packed cup)
¾ cup/175 grams kosher salt
One 2- to 3-pound/1- to 1.5-kilogram salmon fillet in one piece,
    no thicker than 1½ inches/3.5 centimeters, skin on,
    pinbones removed
¼ cup/60 milliliters Pernod
1 fennel bulb, with stalks and leaves, thinly sliced
½ cup/65 grams fennel seeds, toasted (see Note below)
2 tablespoons/20 grams white peppercorns, toasted and
    cracked (see Note below)

1. Mix the sugars and salt well. Sprinkle half of the mixture over the bottom of a nonreactive pan or baking dish just large enough to hold the salmon. Pan size is important, because the fish will release a lot of liquid, forming in effect a highly seasoned brine in which it will cure; and you want the brine to cover as much of the fish as possible. (If you don't have a pan the right size, you can use aluminum foil to wrap the fillet in an enclosed package; the salt won't have enough time to react with the foil, so that is not a problem.) Place the salmon on the salt mixture. Sprinkle both sides of the fish with the Pernod, then cover with the remaining salt mixture. Layer the sliced fennel over the top, followed by the fennel seeds and white peppercorns. Cover with plastic wrap (or enclose completely in the foil).

2. Place a pan on top of the salmon and weight it: A few canned goods will do the trick, as will a brick–try to use 4 to 8 pounds/2 to 4 kilograms. (The idea is to speed up water loss from the salmon by pressing it out, so the more evenly the fish is pressed, the better.) Refrigerate for 48 hours, redistributing the cure ingredients as necessary over the salmon once about midway through the curing. The salmon should be firm to the touch at the thickest part when fully cured. If it still feels raw and squishy, cover and leave in the cure for 24 more hours.

3. When the salmon is fully cured, discard the fennel and spices, rinse it well under cool water, and pat it dry. To store it, wrap in butcher's paper or parchment paper and refrigerate. The salmon will keep for 3 weeks in the refrigerator; rewrap in fresh paper if the paper becomes too wet.

There are many ways to serve cured salmon, but it's commonly sliced translucently thin. This requires a good slicing knife and some practice, but it makes a big difference in flavor and texture. Alternatively, the salmon can be diced for tartare. It can be served with the traditional sieved hard-boiled egg and diced red onion garnish, on toast points or blini with sour cream or crème fraîche. In Scandinavia, some of the juice from the cure may be strained and used to flavor a sauce; a tablespoon would work well mixed with a cup of sour cream and some Dijon mustard. The sliced salmon can also be used in a salad. This particular version works beautifully with a salad of shaved fresh fennel, sliced red onion, and shaved Parmigiano-Reggiano.

Yield: 2 to 2½ pounds/1 to 1.25 kilograms cured salmon

[NOTE: To toast peppercorns, or any whole spices, heat them gently in a small dry skillet until they begin to release their fragrance, a few minutes. They can also be toasted in a 300-degree-F./150-degree-C. oven for 5 minutes or so.]

CURING SALMON

Mise en place for salmon that's going to be cured. The salmon will be rubbed on the top side with plenty of cure, then nestled in the remaining cure and covered with plastic wrap.

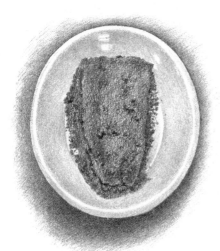

The salmon on the cure. The dry ingredients will soon liquefy as the fish releases moisture and become, in effect, a brine.

A plate or some other weight is placed on top of the curing salmon to help the cure extract moisture from the fish.

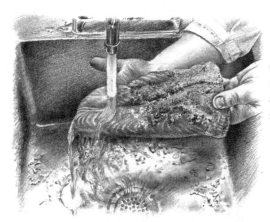

Once the salmon has cured, the fish is rinsed under cold running water to remove all the cure from the surface. Then the salmon is patted dry with paper towels.

If you're smoking the salmon, it should be refrigerated, uncovered, on a rack overnight to allow it to dry and develop a pellicle, a tacky surface the smoke will adhere to. The better the pellicle, the more effective smoking will be.

DUCK PROSCIUTTO

Duck breast is one of the easiest dry-cured items to prepare at home. Duck breasts are widely available, and their size is perfectly manageable. Even a frozen store-bought breast of duck will provide results that will surprise you, especially if you've never cured your own meats, yielding a rich prosciutto-like flavor. You can add additional seasoning here–garlic, juniper, and bay are good–but salt and pepper are really all that's needed for a great duck ham.

This is best sliced, skin side up, on a bias as thin as possible and served like prosciutto, with Parmigiano-Reggiano, arugula, and lemon juice, or with the ham's traditional companion, melon, and a few drops of good balsamic vinegar. It also makes an excellent addition to a charcuterie plate or a canapé. Or arrange a border of slices around a plate, garnish with a mixed green salad with a basic vinaigrette, and you've got an extraordinary course to begin or end a meal.

The type and size of duck aren't critical in this recipe, though if you want an extraordinary duck prosciutto, order magrets, moulard duck breasts, which come from the ducks used for foie gras. They're especially thick and flavorful, in this or any other duck breast preparation.

About 2 cups/450 grams kosher salt, or as needed
1 whole boneless Pekin (Long Island) duck breast, about 1 pound/450 grams, skin on, split
½ teaspoon/1.5 grams freshly ground white pepper
Cheesecloth

1. Put 1 cup/225 grams of the salt in a nonreactive baking pan or dish that will just hold the duck breasts without touching and nestle the duck breasts skin side up on top of the salt (the snugger the fit, the less salt you'll need to use, but be sure that the pieces don't touch each other). Pour enough additional salt over the duck breasts so that the pieces are completely covered. Cover with plastic wrap and refrigerate for 24 hours.

2. Remove the duck from the salt, rinse thoroughly, and pat completely dry with paper towels. The flesh should feel dense, and its color will have deepened. Dust the breasts on both sides with the white pepper.

3. Wrap each breast in a layer of cheesecloth and tie with string. Hang the duck breasts for about 7 days in a cool, humid place (about 50 to 60 degrees F./8 to 15 degrees C. is optimal). The flesh should be stiff but not hard throughout; the color will be a deep rich red. If the breasts still feel squishy (raw) in the center, hang for a day or two longer as needed.

4. Remove the cheesecloth, wrap the duck in plastic wrap, and refrigerate until ready to use. The duck will keep refrigerated for several weeks or more.

Yield: 2 cured half duck breasts

BEEF JERKY

Jerky is an ancient form of preservation used in America and throughout the world. Native Americans took advantage of the technique of drying red meat in strips, and cowboys always had a great need to preserve beef–some still do. A contemporary of ours, a former cowboy who worked the land from Texas to Canada, tending and protecting cattle over huge expanses of territory, often alone for long stretches, says he dried, or jerked, meat whenever he killed anything, such as a deer. When his food ran out, he'd butcher one of the cattle. As that would give him, even if he was traveling with several others, too much meat to keep fresh, he would cut the meat into thin strips, salt it if he had salt, and keep it in a shady cool spot during the day (it would spoil in the sun), then hang it each night. If he was on the move, Robert said, jerky was important because you can carry a lot of it easily, it weighs so much less than fresh meat.

For that reason, jerky remains a great item for long camping trips. The same method works well with venison. Traditionally lean meat was used, because it dried more thoroughly and lasted longer, since even preserved fat can become rancid. Beef round is about as bland and tasteless a cut as you can find, but by curing, seasoning, and drying it, you can transform it into something both superlative and useful.

Beef jerky recipes are as varied as your imagination. Robert liked to season his with a variety of chile peppers. Be sure to trim away all the fat, which could become rancid.

> 2¼ pounds/1 kilogram boneless beef, eye of the round or
> lean round, all fat trimmed away
> 1½ tablespoons/20 grams kosher salt
> 1¾ teaspoons/5 grams garlic powder
> 1¾ teaspoons/5 grams onion powder
> ¼ cup/60 grams finely chopped chipotle peppers packed
> in adobo sauce

1. Cut the beef into strips about ⅛ inch/0.25 centimeter thick and 1 inch/2.5 centimeters wide (length is not critical). In a dish or bowl, combine the remaining ingredients, add the beef, and toss to coat evenly. Cover and refrigerate for 24 hours.

2. Place the strips of beef on a rack set over a baking sheet, so that all sides dry. Turn the oven to 90 degrees F./32 degrees C., put the pan in the oven, and dry the beef for 16 to 20 hours. (If your oven cannot be set so low, try the lowest setting with the door propped open, and check every so often; it may take less time. Depending on the climate and conditions where you live, the beef may even dry well at room temperature.) The beef should be completely dry to the touch, dark, and very stiff.

3. Stored in an airtight container, the jerky will keep for several months or longer at room temperature.

Yield: Approximately 1 pound/450 grams beef jerky

LEMON CONFIT

Lemon confit, or preserved lemon, is a powerful seasoning and a great pantry item to have on hand. A common ingredient in North African and Middle Eastern cuisines, it adds a beguiling lemony-salty brightness to stews, curries, and sauces. It is amazing minced and tossed into a salad, or used to infuse olive oil for a vinaigrette or condiment. And it goes beautifully with chicken, fish, and veal. There may be no purer example of salt's transformative powers than what it alone does to lemon.

Kosher salt to cover (about 2 pounds/1 kilogram)
12 lemons, scrubbed and halved crosswise

1. Pour 1 inch/2.5 centimeters of salt into a lidded nonreactive container just large enough to contain the lemons and salt. An earthenware crock is ideal, but a plastic sherbet container or a wide-mouthed glass jar work fine. A container in which your lemons fit neatly, without being jammed together, will require less salt. Place the lemons in the container, then pour in more salt to cover. They should be completely submerged.

2. Cover and store in a cupboard or other dark place for at least 1 month, preferably 3 months. Once cured, the lemons will keep almost indefinitely in the salt.

3. To use the confit, remove a lemon half, or as many as you need, from the salt and rinse off. The lemon rind will be tan in color. Cut it in half and scrape out the pulp and pith; discard them. Mince or slice the rind. If using it uncooked (in a salad, for instance), blanch it in simmering water for 30 seconds to remove excess salt.

Yield: 24 pieces lemon confit

Brines: The Salt Solution

A brine is simply salty liquid (a dry cure with water). Doesn't sound like much, but in fact, when salt combines with water, its power is magnified. Salt in solution penetrates meat faster, it is the most effective marinade possible because it can flavor the meat down to its center via osmosis, and it results in a juicer finished dish. And it can also be used to preserve meat.

Brines, more so than dry cures, are an excellent way to impart seasoning and aromatic flavors. A brine penetrates a chicken or a pork loin rapidly and completely, bringing with it any flavors you might have added to the salty solution (garlic, onion, tarragon, pepper). Chefs often use brines for pork, chicken, and turkey–the three types of meat that benefit the most from brining–because they result in a uniformly juicy loin or bird that's perfectly seasoned every time.

There's a pleasing contradiction in the brine. Salt, as we know, dehydrates–it draws moisture out of the cells of the meat. Yet while a brine's main effect on meat is to dehydrate it, a brine nevertheless results in a moister, juicier cut, and one that stays juicy even on reheating. The reason for this seeming contradiction, explains Russ Parsons, who has written about the subject both in the *Los Angeles Times* and in his book, *How to Read a French Fry*, is that salt changes the shape of the protein in the meat or bird so that it can actually hold more juice than unbrined meat. Salt allows the protein molecules to expand, to connect more loosely, and thus contain more water within each cell. Salt makes each cell, in essence, plump up.

Because of this effect, roasting a brined chicken or turkey and hitting it at just the right point of doneness is easier than with an unbrined chicken. You can actually overcook it, in fact, and it can still be juicier than a perfectly cooked bird that wasn't brined. The brine seems to allow the breast to withstand the high temperature while the slowpoke legs and thighs continue to cook. At Five Lakes Grill, Brian always brines the pork chops so that no matter who's working the line that day, the end result will be exactly the same, a chop with just the right amount of seasoning and always very, very juicy.

And there's an added benefit that usually goes unmentioned, but is not insignificant. Brined meat has been shown to harbor lower populations of harmful bacteria than unbrined meat, the result of salt's time-tested effect of curing. It dehydrates those living things, either killing them or preventing them from multiplying.

Turkeys, because of their very large, very lean breasts, benefit most from brining. Do you always want to brine a chicken before you roast it? Probably not, because it requires some additional work–and forethought. Also, the texture of the skin is altered by brining, cooking up dark brown and shiny because the brine has taken all the water out of it. It's a lot easier simply to salt the chicken and pop it into a hot oven. But if you like to cook, you will certainly want to have the brining technique (and a basic brine ratio) in your repertoire. (If you're thinking ahead but aren't going to brine the bird, salt your bird a day before cooking it, and it will be even better, the salt having had a chance to penetrate the flesh deeply, as with a brine. Really the only culinary difference between a brine–salt with water–and a dry cure–salt without water–is the ability to introduce a variety of seasonings and flavors in a way that they, and the salt, uniformly surround and penetrate the meat.)

A brine should also include a third component: sugar. Sugar helps to balance out some of the harsher effects of salt and enhances flavor. Finally, because sugar browns so nicely, it helps to give a deep rich color to the surface or skin of roasted meats.

Sugar can usually be used in any of its various forms–white or brown sugar, honey, molasses, dextrose, maple sugar or syrup. Depends what you're making. Dextrose is a refined corn sugar, sometimes called baker's sugar; it is very fine and therefore dissolves easily. For ham or fresh pork, brown sugar or honey has an excellent effect.

A fourth optional but often desirable component is an additional flavoring ingredient–herbs, spices, or aromatic vegetables. As the brine itself penetrates the meat, it brings with it the additional flavors. You might add tarragon, parsley, and thyme to your salted water for an herb-brined turkey or turkey breast. Brian's Five Lakes brine for pork chops includes sage, garlic, and juniper berries. Corned beef uses pickling spices in its brine. If you use aromatic components (which in chefs' vernacular are called *aromats*), it's important to heat your brine first so that the flavors fully infuse the liquid. This also helps to ensure that the salt and sugar completely dissolve and disperse evenly throughout the brine. It's then critical to chill your brine before adding the meat, or you'll be making bad soup–cooking your meat rather than brining it. Adding other aromats in addition to the herbs and spices, such as onion and carrot, to a flavored brine (essentially combining the brining technique with a court-bouillon, or quick stock, technique) is rarely a bad idea if you have the time. In the end, though, it's the salt in the water that works the magic.

- Because brines are such powerful tools, you need to **use them with care**. It's all too easy to infuse meat with too much salt. The saltiness of the meat is determined by two factors:

how salty the brine is and how long the meat stays in the brine. If you leave a 3-pound chicken in our brine for longer than 24 hours, the meat will eventually become too salty. A fat turkey breast, though, needs 24 hours for good brine penetration. As in all seasoning, it's better to undersalt, because you can always add salt if something is underseasoned. If, however, you leave your meat in the brine too long, or suspect it's become too salty, there are ways to fix the problem. To check, cut off a piece, rinse and dry it, and then cook it; it should taste a little too salty, having come from the surface (where the salt concentration is highest immediately after brining), but not unpleasantly so. If your meat does taste too salty, then you need only rethink how you'll handle it. You can immerse it in unsalted water for half the time it was in the brine, and this will reduce its salt content, salt always seeking equilibrium. Or you could braise your meat rather than roast it, cooking it in water or stock with vegetables, and the salt will season the other components.

- We recommend a 5-percent brine: 50 grams of salt per liter of water, or 5 ounces salt per 100 ounces of water. (This ratio, incidentally, also happens to be perfect for blanching green vegetables.) If you're concerned about salt intake, you can reduce the salt and sugar in your brine, but remember that only a small percentage of the salt enters the meat.

- When you remove the meat from the brine, **always discard the brine**. Never reuse it—it's not only diluted from the meat juices, it's also infused with impurities from the meat.

- **Pay attention to the recommended brine time.** Also see the box on brining times on page 60.

- **Allow the meat to rest in the refrigerator after it's been brined,** a couple hours for small items and up to a day for larger cuts. Resting allows the salt within the meat to disperse more evenly. Immediately after it's taken out of the brine, the meat closer to the surface will have a higher salt concentration than the interior will. When a turkey, say, is allowed to rest, the salt seeks equilibrium and continues to migrate until the salt concentration in the cells is uniform throughout, which will result in a uniformly seasoned bird. Resting also allows the exterior to dry, which results in crisper skin.

- **Always chill the brine thoroughly before adding the meat to it.** Chilling a brine can take time. *To reduce the time needed for cooling*, heat all the other brine ingredients in half the required amount of water. Once the salt and sugar are dissolved, add the remaining water, either cold or as ice water.

- **Choose a container that's taller than it is wide when brining larger items,** such as a whole chicken or turkey or a fresh ham, because that will require less liquid to submerge the food.

- **Always brine meat in the refrigerator.**

THE ALL-PURPOSE BRINE

This is the essential brine and can be used for any meat. Adding optional seasonings, depending on the type of meats, will enhance the flavor of the meat.

> 1 gallon/4 liters water
> ³⁄₄ cup/200 grams kosher salt
> ½ cup/125 grams sugar
> Optional seasonings as desired; see below

Combine all the ingredients in a large pot and bring to a simmer, stirring to dissolve the salt and sugar. Remove from the heat and allow to cool to room temperature, then refrigerate until chilled.

> Yield: 1 gallon/4 liters

Basic Fish Brine

> 2 sliced lemons
> ½ cup/60 grams coarsely chopped fresh dill
> ¼ cup/6 cloves coarsely chopped garlic

Follow the All-Purpose Brine recipe (above), adding the lemons, dill, and garlic along with the salt and sugar.

Optional Brine Seasonings

For most brines, any combination of aromatic vegetables—onion, carrot, and celery, roughly chopped or sliced—are a good idea, and the more of them the better. Garlic and peppercorns are excellent additions to most brines, as are the standard herbs, parsley, thyme, and chives, and bay leaf. It's hard to use too many of these ingredients.

Optional Seasonings for Pork

Sage and garlic are excellent (see the recipe on page 64), as are other traditional pork pairings such as juniper berries. Bay leaves and coriander seeds (toasted, of course), as well as brown sugar, also enhance pork. If you are brining the pork to cure it, such as a loin for Canadian bacon, a fresh ham, or even pork belly for bacon, add 1½ ounces/42 grams of pink salt (8 teaspoons) for each gallon of brine.

Optional Seasonings for Poultry

In addition to the standard aromatics mentioned above, tarragon is excellent with chicken. So are lemons; halve and give them a squeeze before plunking them in (see the recipe on page 61).

Optional Seasonings for Fish

Dill and garlic go well with fish, as do lemons; see the variation.

BRINE TIMES

The time you leave a piece of meat in the brine is critical. If you leave it in too long, it can become unpleasantly salty; it's always better to err on the not-long-enough side. Smaller items need only a matter of hours, big items sometimes require a matter of days. Again, brine the item well ahead of cooking it so that it can rest to allow the salt remaining in the flesh to distribute itself evenly, from 2 hours to a day.

Boneless chicken breasts (8 ounces/225 grams): 2 hours

Pork chops, 1½ inches/3.5 centimeters thick: 2 hours

A 2-pound/1-kilogram chicken: 4 to 6 hours

A 3- to 4-pound/1.5- to 2-kilogram chicken: 8 to 12 hours

A boneless turkey breast, 4 inches/10 centimeters thick:
12 to 18 hours

A 4-pound/2-kilogram pork loin: 12 hours

A 10 to 15-pound/4- to 7-kilogram turkey: 24 hours

A turkey over 15 pounds/7 kilograms: 24 to 36 hours

Fish: 1 hour for thin fillets, 6 to 8 hours for fillets or steaks
1 inch/2.5 centimeters thick or more

RECOMMENDED FINISHED TEMPERATURES FOR MEATS

Remember that meat will continue to cook after it's out of the oven (this is called carryover cooking), and its internal temperature will rise five to ten degrees. All meats should be allowed to rest. Roast chicken should rest for at least 15 minutes before serving, and turkey for 20 to 30 minutes.

- Pork: We prefer pork medium-rare to medium, 130 degrees F./54 degrees C., for a finished temperature of 135 to 140 degrees F./57 to 60 degrees C. (In 2011, the USDA reduced the recommended temperature from 165 degrees F./74 degrees C. to 145 degrees F./63 degrees C.)
- Poultry: Poultry should be cooked until the thigh registers 160 degrees F./70 degrees C., for a final temperature of 170 degrees F./76 degrees C. A good way to check if a chicken is done is to tilt it to allow the juices in the cavity to run out. If they are clear, and free of blood, the chicken is done.

HERB-BRINED ROASTED CHICKEN OR TURKEY

Brining a chicken or a turkey is a foolproof way to ensure juicy white meat and completely cooked dark meat. That's especially valuable with a turkey, but it also gives you a little leeway when cooking a chicken. The following recipe will work for a chicken, a boneless turkey breast, or a whole turkey. You may need to double all amounts for the brine ingredients if you're cooking a large turkey, 18 pounds/8 kilograms or more, depending on your brining container. You need a container large enough to hold the turkey completely submerged (check beforehand), but if your vessel is too big, you may need to make more brine to submerge the turkey.

For the brine, it's fine to include the herb stems and the papery garlic and onion skins, if they're free of dirt. When planning this meal, be sure to allow enough time to let the chicken or turkey rest after brining, optimally for a day, but at least for a couple of hours.

The skin of brined birds can become so dehydrated that, with the sugar in the brine, it sometimes crisps quickly as the bird cooks and can become too dark; if you see this happening, place a sheet of aluminum foil loosely over the bird to deflect the direct heat.

THE BRINE
1 gallon/4 liters water
1 cup/225 grams kosher salt
½ cup/125 grams sugar
1 bunch fresh tarragon (about 1 ounce/25 grams)
1 bunch fresh parsley (about 1 ounce/25 grams)
2 bay leaves
1 head garlic, halved horizontally
1 onion, sliced
3 tablespoons/30 grams black peppercorns, lightly crushed
      with the bottom of a sauté pan
2 lemons, halved

One 3- to 5-pound/1.5- to 2.25-kilogram chicken or turkey
      breast

1. Combine all the brine ingredients in a pot large enough to hold the chicken or turkey; give the lemon halves a good squeeze as you add them. Place over high heat and bring to a simmer, stirring to dissolve the salt and sugar. Remove from the heat and let cool to room temperature, then refrigerate the brine until it's chilled.

2. Add the chicken or turkey to the brine. Weight it down with a plate or other object to keep it completely submerged and place in the refrigerator for the appropriate time (see Brine Times, page 60).

3. Remove the chicken or turkey from the brine, rinse well, and pat dry. Let rest uncovered in the refrigerator for 3 to 24 hours.

4. Preheat the oven to 450 degrees F./230 degrees C.

5. Roast the chicken or turkey until it reaches an internal temperature of 160 degrees F./71 degrees C. (the cavity juices will be clear when it's done). Remove from the oven and let rest for at least 15 minutes before serving.

BRINING CHICKEN

The *mise en place* for brining a chicken: chicken, salt and seasonings, cold water, and a container bigger than the bird.

It's critical when brining any meat that it remain completely submerged in the brine. It's almost always necessary to weight the meat down; here a plate is used. Note that if this were for Garlic-sage–Brined Pork Chops (page 64) or any other brine that includes aromatics such as herbs or onion or lemon or garlic, those aromatics would be in this brine liquid.

After brining, the chicken should be refrigerated, uncovered, on a rack so that it can dry and form a pellicle, or tacky surface. The better the pellicle, the better the smoke will adhere to the skin. (Note that this bird is trussed. Trussing helps the bird to cook more evenly, and a bird that's been trussed is much more elegant to look at and to serve. If you don't feel comfortable trussing your own chicken, ask your butcher to do it. Ruhlman.com/2010/07/how-to-truss-a-chicken.)

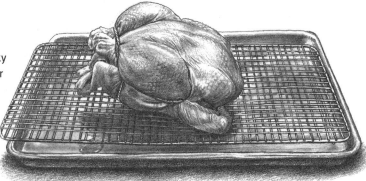

GARLIC-SAGE–BRINED PORK CHOPS

Brian brines a whole bone-in pork loin, then cuts it into thick chops, ties each chop to keep the meat tight and uniform while cooking, and grills them (he simplifies the method here). Grilling adds smoky flavors, but these can be pan-roasted as well. Brian pairs the chops with a mustard-horseradish sauce, herbed spaetzle, and creamed savoy cabbage. But, the chops would work well with just about any starch–roasted potatoes, polenta, pasta, rice–and vegetable, though winter root vegetables are particularly suited to it. And any kind of mustard sauce tends to work great with pork.

THE BRINE

2 quarts/2 liters water

½ cup/125 grams kosher salt

¾ packed cup/135 grams dark brown sugar

4 packed tablespoons/24 grams fresh sage leaves

1 tablespoon/8 grams juniper berries, crushed with the
  side of a knife

2 garlic cloves, lightly smashed with the side of a knife

1 tablespoon/10 grams freshly ground black pepper

Four 1½-inch/3.5-centimeter-thick bone-in pork chops,
  12 ounces/350 grams each (or 1 bone-in pork loin;
  see the headnote)

1. Combine all the brine ingredients in a pot large enough to hold the pork and bring to a simmer, stirring to dissolve the salt and sugar. Remove from the heat and let cool completely, then refrigerate until cold.

2. Add the pork chops to the brine and refrigerate them for 2 hours weighing them down if necessary. (If brining a larger piece of pork, increase the brine time to 6 hours for thick chops and 12 hours for an entire bone-in loin, see page 60 for brine times.)

3. Remove the pork from the brine, rinse under cold water, and pat dry with paper towels. Refrigerate, uncovered, for at least an hour, or up to a day.

4. *To grill the chops:* Prepare a hot fire in one side of a grill. Sear the chops over the hot coals on both sides, then move them to the side away from the coals, cover the grill, and finish cooking over indirect heat to an internal temperature of 130 to 140 degrees F./54 to 60 degrees C. about 10 more minutes.

To pan-roast the chops: Preheat the oven to 350 degrees F./175 degrees C. Heat an ovenproof sauté pan over high heat, add a film of canola oil, and sear both sides of the chops. Place the pan in the hot oven and roast until the chops reach an internal temperature of 130 to 140 degrees F./54 to 60 degrees C., 10 to 15 minutes.

5. Regardless of the cooking method, let the pork rest for 5 minutes before serving.

Yield: 4 servings

CORNED BEEF

Making your own corned beef is especially satisfying because it's so easy—and so inexpensive compared to commercial corned beef. It's also a pleasure to have a hand in what is an extraordinary transformation of a cheap cut of meat. We love simple braised brisket, like the Belgian stew carbonnade, cooked slowly in beer and onions, but to cause the metamorphosis from brisket to delicious corned beef is a different pleasure altogether. It becomes firmer, it takes on the delicious cured flavor, and, while it's excellent for sandwiches, it can make an elegant main course for a full meal, served with, say, sautéed blanched cabbage or Brussels sprouts with a mustard vinaigrette and boiled potatoes. When making a meal of it, include an onion and carrot and other aromatics in your poaching liquid and then spoon it, strained, like a jus or a broth over the corned beef.

THE BRINE
1 gallon/4 liters water
2 cups/450 grams kosher salt
½ cup/100 grams sugar
1 ounce/25 grams pink salt (4 teaspoons)
3 garlic cloves, minced
2 tablespoons/20 grams Pickling Spice (page 68
        or store-bought)

One 5-pound/2.25-kilogram well-marbled (first-cut) beef
        brisket
2 tablespoons/20 grams Pickling Spice

1. Combine all the brine ingredients in a pot large enough to hold the brisket comfortably. Bring to a simmer, stirring until the salt and sugar are dissolved. Remove the pot from heat and allow to cool to room temperature, then refrigerate the brine until it's completely chilled.

2. Place the brisket in the brine and weight it down with a plate to keep it submerged. Refrigerate for 5 days.

3. Remove the brisket from the brine and rinse it thoroughly under cool running water. (Resting is not required here because the distribution of the brine will continue in the long, slow cooking process.)

4. Place the brisket in a pot just large enough to hold it and add enough water to cover the meat. Add the remaining pickling spice and bring to a boil, then reduce the heat, cover, and simmer gently for about 3 hours, or until the brisket is fork-tender (there should always be enough water to cover the brisket; replenish the water if it gets too low).

5. Remove the corned beef from the cooking liquid (which can be used to moisten the meat and vegetables, if that is what you're serving). Slice the beef and serve warm, or cool, then wrap and refrigerate until you're ready to serve, or for up to a week.

Yield: 4½ pounds/2 kilograms corned beef; 8 to 10 servings

Pickled Vegetables

You don't need vinegar to pickle vegetables. Vegetables submerged in a mild brine and left at cool room temperature (65 to 70 degrees F./18 to 21 degrees C.) will, in a week's time, take on an appealingly sour flavor as the desirable bacteria create lactic acid–thus, the pickle. The developing acidic environment prevents the growth of harmful bacteria that would otherwise cause spoilage. The key to the fermentation is a mild brine that is just salty enough to kill that bacteria but not so salty as to be unpalatable or to prevent the *Lactobacillus* bacteria, present in the atmosphere, from producing the necessary acid.

Our friend Michael Pardus, a chef-instructor at the Culinary Institute of America in Hyde Park, New York, often finds himself with an abundance of fresh vegetables in late summer and has been experimenting with natural pickles like these so that the exquisite Hudson Valley produce can be enjoyed year-round. He's learned that a 5-percent brine (50 grams of salt per liter, or a little less than 2 ounces, about ¼ cup, per quart) is perfect for just about any vegetable. Hearty root vegetables such as carrots make excellent pickles.

"I've found that carrots, cucumbers, small turnips, radishes, green tomatoes, peppers–

both hot and sweet—onions, green beans, mushrooms, and eggplants all work well," Michael says.

The quality of the vegetables going in (especially with cukes) makes a big difference in the success of the pickle. Also, it can be difficult to pickle in warm weather, because temperatures above 75 degrees F./23 degrees C. can allow bad bacteria to take over. You can pickle in the refrigerator, but it can take three times as long, depending on the density of the vegetable. The other critical matter is ensuring that all the vegetables are kept submerged. Otherwise, Michael says, "evil molds and slimy things start to grow." Aromatics, garlic, herbs, and chiles all add flavor and, presumably, more helpful bacteria for a livelier pickle.

A classic sauerkraut is a pure form of pickling: sliced cabbage is salted, weighted and covered, and left in a cool place to ferment, the cabbage releasing enough water to create its own brine.

THE NATURAL PICKLE

Here is the basic method for pickling by fermentation. It's a simple process that requires no special tools other than the right-sized container. The precise salt-to-water ratio is 50 grams per liter, for a 5-percent salt solution (one of the many examples of why the metric system is best in the kitchen; we've also given amounts for our less-efficient system).

> 1¾ tablespoons/50 grams kosher salt
> Optional but recommended seasonings: aromatics such as
>   garlic, ginger, and chiles, peppercorns, or fresh herbs
>   such as thyme or tarragon, or Pickling Spice (below)
> 4¼ cups/1 liter water
> 8 ounces/225 grams trimmed (and peeled, if appropriate)
>   vegetables (carrots, beans, onions—see the
>   suggestions above)

1. Combine the salt, the seasonings, if you're using them, and the water in a saucepan and bring to a simmer, stirring to dissolve the salt. Remove the brine from the heat and let cool to room temperature.

2. Place the vegetables in a clean jar that will hold them comfortably. Pour enough brine over the vegetables to cover them. Fill a small plastic bag with more brine and press this on top of the vegetables to keep them submerged, or press a piece of plastic wrap down on top of the vegetables and pour more brine onto it. It's essential that the vegetables remain completely covered.

3. Place the jar in a cool spot to ferment for 7 days. (Any temperature above 75 degrees F./23 degrees C. or so will encourage the less benevolent bacteria to take over. See the headnote for more information.

4. After 7 days, taste the vegetables. They should retain their crunch and have a mild salty-sour taste (not sour like vinegar). If you would like a stronger sour flavor, cover and let the vegetables sit for 3 more days, or until the desired sourness is reached.

5. To store the pickled vegetables, remove them from the brine and place in a clean jar or other container. Strain the brine into a pan and bring to a boil (this is a precaution against any harmful bacteria that might still be present). Remove from the heat and allow to cool to room temperature.

6. Pour the cooled brine over the vegetables, cover, and refrigerate. (This is an exception to the never reuse brine rule; here it's reused for storage, not for brining.) The vegetables will keep indefinitely in the refrigerator.

Yield: 8 ounces/225 grams pickled vegetables

Pickling Spice

Commercial versions of pickling spice, available in the spice section of the grocery store, are acceptable, but this version is a little sweeter smelling and doesn't have the pungent bay leaf aroma that dominates most store-bought brands. It's a fabulous mixture. Make big batches and freeze it to have on hand.

2 tablespoons/20 grams black peppercorns

2 tablespoons/20 grams mustard seeds

2 tablespoons/20 grams coriander seeds

2 tablespoons/12 grams hot red pepper flakes

2 tablespoons/14 grams allspice berries

1 tablespoon/8 grams ground mace

2 small cinnamon sticks, crushed or broken into pieces

24 bay leaves, crumbled

2 tablespoons/6 grams whole cloves

1 tablespoon/8 grams ground ginger

1. Lightly toast the peppercorns, mustard seeds, and coriander seeds in a small dry pan, then smash with the side of a knife just to crack them.

2. Combine the cracked spices with the remaining ingredients, mixing well. Store in a tightly sealed plastic container or glass jar.

Yield: 1 cup/125 grams

TRADITIONAL DILL PICKLES

The quality of the vegetable is especially important here. If you start with mediocre grocery store baby cukes, you're not likely to produce a crisp dill pickle. We recommend that you only pickle little cucumbers in season, when they are abundant, preferably ones you grow yourself or find at a local growers' market. If you start with fresh-from-the garden cucumbers, the rest is simple.

THE BRINE
5 tablespoons/65 grams kosher salt
1 teaspoon/2 grams dill seeds
½ cup/125 milliliters white wine vinegar
1 teaspoon/3 grams black peppercorns
1 tablespoon/10 grams Pickling Spice (page 68 or store-bought)
5 cups/1.25 liters water

1 bunch fresh dill
10 pickling or baby cucumbers (about 1 pound/450 grams),
     washed

1. Combine all the brine ingredients in a small pot, bring to a boil, and boil for 3 minutes. Remove from the heat and let cool completely.

2. Place the dill and cucumbers in a jar or other nonreactive container just large enough to hold them and the brine and pour the brine over them. Allow to pickle in the refrigerator for 3 weeks, or until they are sour and thoroughly flavored by the herbs and spices. The pickles will keep a month or more in the refrigerator.

Yield: 10 pickles

HOME-CURED SAUERKRAUT

Curing your own cabbage results in delicious bright sauerkraut and a fresh tart flavor that's far superior to the bagged version at your grocery store. It is so easy to do it's well worth your while. The only caveat is that it's a two-week fermentation, so plan ahead.

You can use any kind of cabbage, but ordinary green cabbage results in the deepest flavor and sturdy texture. After the cabbage is cured, the best way to serve the sauerkraut is braised in half pickling liquid and half chicken stock or water (use less or more stock or water to decrease or increase the acidity). Bring to a simmer in an ovenproof sauté pan on the stovetop then move it to a 300-degree-F./150-degree-C. oven for up to 30 minutes, until ready to serve. Add a bay leaf or other aromatic seasoning as you wish.

THE BRINE
17 cups/4 liters water
¾ cup plus 2 tablespoons/200 grams kosher salt

1 green cabbage, about 3 pounds/1.5 kilograms, thinly sliced or shredded

1. Combine the water and salt in a small pot and bring to a simmer, stirring to dissolve the salt. Remove from the heat and let cool, then chill.

2. Combine the cabbage and brine in nonreactive container. Cover the cabbage with a piece of cheesecloth or a clean kitchen towel, then weight the cabbage and cloth down with a plate, pressing the plate down so that the cabbage is completely submerged. Cover loosely with plastic wrap and set in a cool place for 2 weeks (no hotter than 70 to 75 degrees F./21 to 23 degrees C., or nonbeneficial bacteria can begin to thrive).

3. Drain the cabbage, reserving the brining liquid; the cabbage should have a pleasant sour-salty flavor and although its green color will have paled, it should still be crunchy. Strain the brining liquid into a pot and cover and refrigerate the cabbage. Bring the brining liquid to a boil. Remove from the heat and let cool to room temperature, then chill.

4. Pour enough of the cold brine over the sauerkraut to cover it completely; discard the extra brine. Store, covered, in the refrigerator for up to 3 weeks.

Yield: 1 quart/1 liter

3.

꙳꙳꙳

SMOKE:
THE EXOTIC SEASONING

. . . .

Curing with salt and smoking go hand in hand. Smoking foods was indeed once an aid in the preservation of meat and fish, but in contemporary cooking it is fundamentally a matter of enhancing flavor and color in substantial ways. Is it pretension on a chef's menu to list "Apple-wood Smoked Bacon"? No: the smoke from the wood of an apple tree or almost any fruit tree produces a sweeter, milder result than a harder wood such as hickory, with its powerfully flavored wood smoke. It matters. Smoke a delicate fish over pear wood, and you'll understand.

In this chapter we discuss smoking fish, meat, vegetables, and nuts.

We smoke foods to give them a great flavor. Smoked meat and fish also take on an appetizing caramel-brown hue. Hot dogs are brown, not pasty looking, because they're smoked. While the smoke coating does have some preservative effects by making the surface of the meat acidic, thereby discouraging the growth of unwanted microorganisms and bacteria, smoke is not used to preserve foods the way drying and salting are. Smoking may have become part of the charcutier's trade because of its initial preservative nature, but we continue to smoke food because of the fine color and flavor it gives to dried and cooked foods, and especially to pork.

Smoke is flavor. It's why we love barbecued ribs, chicken on the grill, burgers cooked over open flame. Smoke is what gives bacon its depth. It's the reason smoked ham hocks are so good with beans or long-simmered greens. Cure salmon in your refrigerator, then smoke it, and you will have transformed it into something truly special. Jalapeño peppers, when smoked, become chipotle peppers, one of the great seasoning elements of Southwestern cuisine. Smoke not only elevates a ham, in many cases the type of smoke used determines the kind of ham it is and the regional nuances that distinguish it. Was it smoked over American hickory and apple wood, traditional woods for the American hams, or over the beech and juniper of Westphalia, Germany? Smoke can describe a culinary tradition and the spirit of the *terroir*.

The smoking environment may be hot, in which case it cooks the meat or fish while enhancing its flavor (as with Canadian bacon), or it may be cold, so the food remains uncooked but takes on a smoky flavor (as with smoked salmon). Smoking at or below 100 degrees F./37 degrees C. is cold-smoking; smoking at between 150 and 200 degrees F./65 and 93 degrees C. is hot-smoking. Meat or sausages that are hot-smoked cook gently for a long time while being flavored by the smoke. They can then be eaten immediately or chilled and later reheated. Pan-smoking (smoking on your stovetop) and smoke-roasting (as in a cylindrical smoker or barbecue grill) occur at temperatures of 300 degrees F./150 degrees C. or more.

Salmon is typically cold-smoked, ideally at a temperature below 90 degrees F./32 degrees C.; if the smoke were hotter, it would cook the fish and drastically change its texture. Some dry-cured sausages, such as pepperoni and Spanish chorizo, are cold-smoked before being hung to dry. Smoked kielbasa and other hot-smoked sausages are hung in the smoker until fully cooked.

There are varying degrees of smoke and temperature, but the basics remain the same:

- We smoke food primarily to make it taste better (smoking has negligible preservative effects).
- We also smoke food to give it a rich color; smoke results in an appetizing appearance.
- The level of heat defines the type of smoking. Cold-smoking does not cook the food; hot-smoking cooks it gently and slowly; and smoke-roasting and pan-smoking cook the food as if it were in a hot smoke-filled oven.
- The longer the meat is smoked, the deeper the flavor and the color will be.

The venerable kitchen rationalist Harold McGee writes: "Smoke's usefulness results from its chemical complexity. It contains many hundreds of compounds, some of which kill or inhibit microbes, some of which retard fat oxidation and the development of rancid flavors, and some of which add an appealing flavor of their own."

The composition of smoke depends, of course, on the substance you're burning. When smoking food over wood, it's critical to use only hardwoods (hickory, maple, fruitwoods). Avoid soft woods (such as pine), heavy-sap-producing wood, green wood, and any treated wood; these can release a sometimes-harmful resin and their smoke coats the food with an unpleasant flavor. Hickory, perhaps the most common choice for smoking, has a strong, smoky flavor and gives a rich amber color, suitable for hearty meats and sausages. Fruitwoods are preferable to harder woods for their mild sweetness. Pear is very mild and gives a light color, making it ideal for delicate fish, such as whitefish. Cherry is a favorite in Michigan, where the trees are abundant—Brian likes to hot-smoke duck breast over cherry. And the pairing of applewood smoke and bacon is so felicitous it's become almost commonplace. But hardwoods do not provide the only smoke beneficial to food: herb branches and tea leaves give off tasty smoke as well.

Home-Smoking

Home cooks can smoke their own food, but results depend on the equipment. You can certainly smoke on your stovetop with a pot or roasting pan, a rack to fit inside, and some sort of cover—and an excellent exhaust system. (You could even use a pot with a steamer insert.) You can smoke on a covered grill by adding hardwood to low coals and keeping the food off to the side, away from direct heat. All kinds of stovetop and outdoor smokers are available today, and these are all hot-smoking devices. They cook while they smoke, which limits the time you can keep the food in the smoke.

Smokers that enable you to smoke at low temperatures generate the smoke outside the smoke box. Most smokers that allow you to adjust the heat are expensive, in the thousands of dollars range, and commercial smokers that allow for cold-smoking cost even more. There are many different smoking options, from big box smokers that provide continuous smoke, such as the Bradley Smoker, to Weber grill smoker inserts, to Kamado-style earthenware grills, such as the Big Green Egg, which is pricey but a fabulous way to smoke bacon and pastrami and other big whole muscles.

True cold-smoking is difficult to do without the proper equipment or a purpose-built smokehouse and smoke pit. Placing a tray of ice between the meat and the smoke source is one way to keep the smoke cool longer. Professional smokehouses that include some sort of refrigeration device and do all the work for you cost as much as a car.

So smoking for the home cook without professional equipment takes some work and often ingenuity. It's possible to smoke for long periods on a grill with a little effort. Bruce Aidells, the San Francisco–area sausage king, writing in *Gourmet* magazine ("Making Bacon," June 2002, p. 72), describes a method whereby he puts a few burning coals into a pie pan filled with wood chips or dust and sets it in a kettle grill. He then places a brine-cured pork loin inside the grill and smokes the pork for six to eight hours. This requires continual maintenance of the smoke as the coals burn out, but the resulting Canadian bacon is very good. If you brine a pork loin using the All-Purpose Brine (page 59), including 2 teaspoons/12 grams of pink salt in the brine, and then smoke it, you'll have Canadian bacon. This method of smoking is also a perfectly acceptable way to smoke your own pork belly for traditional bacon. In the same way that a pork loin (or a pork shoulder, for that matter) takes on a dark color and a rich smoky flavor, so too does cured pork belly. Also, an item can be smoked on a grill then finished in a low oven.

HOT-SMOKING, COLD-SMOKING: DEFINITION AND METHOD

All of the following recipes, and many in the following chapter, involve smoke. Most recipes instruct the cook to "hot-smoke" a food.

To hot-smoke means to cook at or above 150 degrees F./65 degrees C. in a smoker. The temperature we recommend for hot-smoking is 180 degrees F./82 degrees C. for sausages (because of their higher fat content) and 200 degrees F./93 degrees C. for whole cuts, allowing for slow cooking and maximum smoke. If you have a smoker with a heat control, hot-smoke all these recipes at 200 degrees F./93 degrees C. unless otherwise specified.

If you don't have a smoker, and are relying on ingenuity, then "hot-smoke" simply means smoking the item as you wish until its internal temperature reaches the desired temperature, measured on an instant-read thermometer (see page 28 for notes on thermometers). To smoke bacon, for instance, you might set five or six burning coals in a pan of hickory sawdust, set the pork belly on the rack, and cover the grill: add a few more coals after an hour. Once you see that you have good color on the bacon, you might finish it in a 200-degree-F./93-degree-C. oven, cooking it to the final temperature. Or, if you have a basic kettle smoker, you might simply make the lowest fire possible and cook the bacon entirely in this smoker, removing it when it reaches 150 degrees F./65 degrees C.

Cold-smoking is defined by a temperature of less than 100 degrees F./37 degrees C. and is difficult to achieve without the proper equipment. But there are new devices and how-to videos available with a quick Internet search.

When a recipe here calls for cold-smoking, it assumes that you have a reliable smoke box that can stay below 100 degrees F./37 degrees C. indefinitely. If you don't, we don't recommend cold-smoking food. To cold-smoke the food, place it in the smoke box for the recommended time, making sure the temperature doesn't rise, ideally, to above 90 degrees F./32 degrees C., and certainly no higher than 100 degrees F./37 degrees C.

If you're the sort of cook who likes to improvise and jury-rig, tend fires, manage smoke, regulate the heat, and generally spend a lot of time hanging out with your food, you'll have no problem with smoking. If that sounds like a headache to you, then just stick to conventional forms of smoking; that is, hot-smoking on a charcoal grill. Even if you have a good smoker, smoking food at lower temperatures takes care and attention.

Smoking and Food Safety

Most recipes involving smoking require pink salt, or sodium nitrite, as an insurance against the possibility of botulism poisoning. The spores that can produce the deadly nerve toxin botulism tend to thrive in smoking conditions (low temperatures over long periods, and the low-oxygen environment inside the smokebox). So in most instances of smoking we recommend using pink salt. Food that is smoke-roasted, however–that goes from the refrigerator into a hot smoker (300 degrees F./150 degrees C. or more)–does not require pink salt. (For a more detailed discussion of nitrites, nitrates, and botulism, see pages 174–176.)

As with foods high in fat and cured foods, smoked cured foods should be eaten in moderation. Smoke is composed of many wonderful compounds but some harmful ones too. Again, Harold McGee: "Prominent among these are the polycyclic aromatic hydrocarbons, or PAHs, which are proven carcinogens and are formed from all of the wood components in increasing amounts as the temperature is raised."

The All-Important Pellicle

Of smoking basics, the only issue that isn't a matter of common sense is the importance of allowing the food to dry long enough before smoking to form a *pellicle*, a tacky surface that the smoke will stick to. (This is especially noticeable with salmon, which develops a distinctly tacky feel when dried.) If you put damp meat or sausage into a smoker, it won't pick up the smoke as effectively as it would if dried uncovered in the refrigerator overnight. Yes, food will still pick up smoke if you don't give it a chance to develop a pellicle, but the end results will be superior if you do.

Little House in the Big Woods

One of our favorite scenes in Laura Ingalls Wilder's classic series comes in the first pages of the first book, *Little House in the Big Woods*. Laura describes Pa hanging strips of venison inside a length of hollow tree and building a fire of moss and bark and green hickory chips inside to smoke the meat, which had been salted for a few days. After several days in the smoke, Ma would remove the meat, wrap it in paper, and store it in the attic. Pa, Laura writes, knew meat thus prepared would keep anywhere in any weather.

Indeed, the whole of that first chapter is a primer in food preservation. The Ingalls family salted fish and kept it in barrels. They stored root vegetables in a cellar, braided and hung onions, stockpiled gourds in the attic. They kept cheeses all winter in the pantry. They fattened a pig for slaughter in the late fall, and whole pages are devoted to how they used it. Hams and shoulders were salted and smoked, lard was rendered and stored in jars (the cracklings reserved to flavor johnnycakes), the head boiled till the meat was meltingly tender and mixed with the "pot-liquor," then cooled into sliceable headcheese. Salt pork was stored in a keg in the shed, and all the little leftover scraps of meat Ma chopped fine and seasoned with dried sage from the garden, then rolled into balls for sausage (the balls would stay frozen in a pan in the shed "and be good to eat all winter").

There's a reason that food preservation opens this durable multivolume saga: It was the most important thing settlers on the frontier did. To survive a Wisconsin winter in the late 1800s, preserving food had to be a matter of course. The early settlers knew how to preserve food—or they simply didn't last very long.

Not a single recipe that follows is required for your survival, but they all taste really good and are fun to make. Here, of course, is smoked bacon, bacon in its most familiar form, but there are also other great pork preparations defined by smoke, such as Canadian bacon (smoked loin), hocks, and the regional specialty from Cajun cuisine, tasso ham, made from slices of shoulder, as well as poultry, fish, even chile peppers and nuts.

HERB-BRINED SMOKED TURKEY BREAST

Turkey is an underused bird, probably because it's been debased by an industry committed to growing turkeys that have enormous quantities of white meat, or perhaps it's simply our holiday conditioning. But when treated with care, a turkey is a great meat to cook at any time of the year. One action we can take, besides buying turkeys from small growers, is to give the meat extra moisture and flavor by brining and then smoking the bird. The curing salt in the brine will give the meat the customary cured flavor. Turkey takes smoke beautifully (try grilling a turkey, and you'll see immediately). The herbs can be varied according to your tastes; their flavor here is secondary to that of the brine and smoke.

This recipe is for a bone-in turkey breast, but you can brine, then smoke a whole turkey (or chicken, for that matter), though you may need to double the brine recipe, depending on the size of your bird and your stockpot. I have various-sized stockpots and have even used a $5 Home Depot bucket when brining a big bird or ham. When you're brining a turkey breast, it's best to use one still on the bone; the bones help retain the shape and moisture and results in deeper, more succulent flavor.

THE BRINE

1 gallon/4 liters water

1½ cups/350 grams kosher salt

½ cup/125 grams sugar

6 teaspoons/42 grams pink salt

2 bunches fresh tarragon

5 garlic cloves, crushed with the side of a knife

1 bay leaf

2 tablespoons/20 grams black peppercorns

1 whole turkey breast, 12 pounds/5.5 kilograms

1. Combine all the brine ingredients in a pot large enough to hold the turkey and bring to a simmer, stirring to dissolve the salt and sugar. Remove from the heat and allow to cool, then refrigerate until chilled.

2. Place the turkey in the brine, and weight it with a plate to keep it submerged. Refrigerate it for 48 hours.

3. Remove the turkey from the brine, rinse it under cold water, and pat dry with paper towels; discard the brine. Refrigerate, uncovered, for at least 2 hours, and as long as a day.

4. Hot-smoke the turkey (see page 74) at 200 degrees F./93 degrees C. to an internal temperature of 160 degrees F./71 degrees C.

Yield: 6 to 8 servings

WHISKEY-GLAZED SMOKED CHICKEN

Here two techniques, glazing and smoking, result in a rich mahogany-colored skin and a sweet tangy flavor from the bourbon, sugar, and cayenne. The technique is simple: the glaze is cooked to a light syrupy consistency and brushed on halfway through the smoking, and then again at the end. Serve hot or cold. It's not always easy to make cold chicken look really appetizing, but this is one sure way to do it. It's excellent served on lettuces with a vinaigrette or with a sharp potato salad.

THE BRINE
1 gallon/4 liters water
1½ cups/350 grams kosher salt
½ cup/125 grams sugar
6 teaspoons/42 grams pink salt

1 chicken, 3 to 4 pounds/1.25 to 2 kilograms

THE GLAZE
1 cup/250 milliliters bourbon or wild turkey
½ cup/125 grams maple sugar or ½ cup/125 milliliters
        maple syrup
¼ packed cup/50 grams dark brown sugar
Pinch of cayenne pepper

1. Combine all the brine ingredients in a pot large enough to hold the chicken and bring to a simmer, stirring to dissolve the salt and sugar. Remove from the heat and allow to cool to room temperature, then refrigerate until chilled.

2. Remove any giblets from the cavity and truss the chicken, then submerge it in the brine. Weight it down with a plate to keep it submerged (see the illustrations on page 63), and refrigerate for 18 hours.

3. Remove the chicken from the brine (discard the brine), rinse under cool water, and pat dry. Refrigerate, uncovered, for at least 3 to 4 hours, and up to a day.

4. Combine the glaze ingredients in a small heavy-bottomed pan and bring to a simmer over high heat, stirring to dissolve the sugar. Lower the heat and gently simmer until thick and syrupy and reduced to about 1 cup/250 milliliters. Remove from the heat and let cool.

5. Begin hot-smoking the chicken (see page 74). Halfway through the smoking–about 1½ hours if you're smoking at about 200 degrees F./93 degrees C.–remove the chicken from the smoker and brush all over with half the glaze. Return the chicken to the smoker and continue cooking until a thermometer inserted into the thigh registers 160 degrees F./71 degrees C.

6. Remove the chicken from the smoker and brush all over with the remaining glaze. It should have a deep brown lacquered color.

7. Serve hot, or cool and refrigerate, covered, until ready to serve.

Yield: 4 servings

HOT-SMOKED DUCK HAM

This preparation, in which duck breasts are brined and then smoked until fully cooked, is reminiscent of ham in flavor and in its ratio of fat to meat. The pink salt keeps the meat pink and gives it a cured flavor, enhancing the smoky, rich, meaty duck flavor. Although you could serve it warm, this is even better when cold, and it is superb as part of a charcuterie plate or sliced and served on greens with a vinaigrette. It pairs especially well with blue cheeses, cherries, and walnuts.

THE BRINE
2 quarts/2 liters water
¾ cup/175 grams kosher salt
¼ cup/50 grams sugar
3 teaspoons/20 grams pink salt
½ cup/125 grams maple sugar or ½ cup/125 milliliters
    maple syrup
½ cup/125 milliliters Madeira
1 bunch fresh thyme
2 bay leaves
1 tablespoon/8 grams juniper berries
1 tablespoon/6 grams chopped sage

6 whole boneless Pekin (Long Island) duck breasts,
    about 6 pounds/3 kilograms, skin on

1. Combine all the brine ingredients in a large pot, place over medium-high heat, and bring to a simmer, stirring to dissolve the salt and sugar. Remove from the heat and allow to cool to room temperature, then refrigerate until completely chilled.

2. Add the duck breasts to the brine and weight them down with a plate to keep them submerged. Refrigerate for 8 to 12 hours.

3. Rinse the breasts under cold water and pat them dry. Refrigerate them on a rack set over a plate, uncovered, for at least 4 hours, and up to 24 hours.

4. Hot-smoke the duck (see page 74) to an internal temperature of 160 degrees F./71 degrees C., about 2½ hours. Refrigerate until ready to serve.

Yield: 6 smoked duck breasts

MAPLE-CURED SMOKED BACON

Today, when people no longer need to preserve food to survive, this recipe is a powerful reminder of America's rich culinary history. Likely made popular by English settlers in the seventeenth and eighteenth centuries (all manner of cured pork sides were, writes Alan Davidson in *The Oxford Companion to Food*, "peculiarly a product of the British Isles"), cured or smoked pork has long been a part of our cooking, essential in regional specialties from New England chowders to Southern succotash. Making your own bacon embodies all the reasons we should take the time to do it at home. There may be no better flavor than good bacon, and even if you only have a charcoal grill, you can achieve excellent results.

Many small producers make excellent versions of bacon in this country, varying with time of the cure and the seasonings used. This recipe is for a sweeter bacon. There should be some sugar or sweetness to balance the salt, but if you prefer a more savory taste, omit the maple syrup. If you like black pepper, add it to the cure. Seasonings can vary infinitely, but it is the curing and the smoke that make bacon one of the greatest flavors on earth.

THE CURE
¼ cup/50 grams kosher salt
2 teaspoons/14 grams pink salt
¼ cup/50 grams maple sugar or packed dark brown sugar
¼ cup/60 milliliters maple syrup

One 5-pound/2.25-kilogram slab pork belly, skin on

1. Combine the salt, pink salt, and sugar in a bowl and mix so that the ingredients are evenly distributed. Add the syrup and stir to combine.

2. Rub the cure mixture over the entire surface of the belly. Place skin side down in a 2-gallon Ziploc bag or a nonreactive container just slightly bigger than the meat. (The pork will release water into the salt mixture, creating a brine; it's important that the meat keep in contact with this liquid throughout the curing process.)

3. Refrigerate, turning the belly and redistributing the cure every other day, for 7 days, until the meat is firm to the touch.

4. Remove the belly from the cure, rinse it thoroughly, and pat it dry. Place it on a rack set over a baking sheet tray and dry in the refrigerator, uncovered, for 12 to 24 hours.

5. Hot-smoke the pork belly (see page 74) to an internal temperature of 150 degrees F./65 degrees C., about 3 hours. Let cool slightly, and when the belly is cool enough to handle but still warm, cut the skin off by sliding a sharp knife between the fat and the skin, leaving as much fat on the bacon as possible. (Discard the skin or cut it into pieces and save to add to soups, stews or beans, as you would a smoked ham hock.)

6. Let the bacon cool, then wrap in plastic and refrigerate or freeze it until ready to use.

Yield: 4 pounds/2 kilograms smoked slab bacon

A slab of pork belly should have equal proportions of meat and fat. This piece has been squared off and is ready for the cure.

To cure bacon, the salts, sugars, and spices are mixed and spread all over the meat. The bacon can be cured in a pan or in a 2-gallon Ziploc bag.

SMOKED HAM HOCKS

Anytime you might add a chunk of ham to your pot of beans, greens, or soup, the smoke-cured effect of these ham hocks would be equally, if not more, welcome. Hocks freeze beautifully, keeping for months if well wrapped.

A ham hock is not typically eaten by itself. Rather, the hocks can be braised and the meat separated from them to make smoked pork rillettes, for example, or a pig's feet dish. Or shred the meat and mix it into polenta for a superlative take on scrapple.

THE BRINE
1 gallon/4 liters water
1½ cups/350 grams kosher salt
1 cup/225 grams sugar
6 teaspoons/42 grams pink salt

8 fresh ham hocks (about 8 pounds/3.5 kilograms total)

1. Combine all the brine ingredients in a pot large enough to hold the ham hocks and bring to a simmer, stirring until the salt and sugar are dissolved. Remove the brine from the heat and let cool, then refrigerate until thoroughly chilled.

2. Add the hocks to the brine and weight with a plate to keep them submerged. Refrigerate for 3 days.

3. Remove the hocks from the brine (discard the brine), rinse well, and pat dry. Refrigerate on a rack set over a plate or a tray, uncovered for 8 to 24 hours.

4. Hot-smoke the hocks (see page 74) to an internal temperature of 150 degrees F./65 degrees C.

Yield: 8 smoked ham hocks

TASSO HAM

Tasso ham is a Cajun preparation of pork shoulder and one of the easiest hams to cure. The shoulder is sliced into slabs, quickly cured, heavily spiced, and then smoked. The hot and savory spices are seasonings prominent along the Louisiana bayou, complex flavors brought there by the French Canadian settlers more than three hundred years ago. Shoulder is a tough cut and so needs long, gentle cooking.

Tasso is not eaten on its own but rather appears as a component in other dishes, most commonly jambalaya or gumbo. A gumbo may be good without chunks of tasso ham that have stewed in it for hours, but it's not as good as it can be. Tasso can be used in any dish the way ham or ham hocks are used. Stewed with mustard greens, or with any bean dish or legume, it becomes tender and succulent. You could even dice it fine and heat it in some oil with minced shallot, then add some spinach and sauté it.

"SALT BOX METHOD": Curing Tasso Ham

The "salt box method" of curing a piece of meat essentially means dredging the meat in a salt cure so that all sides are evenly coated.

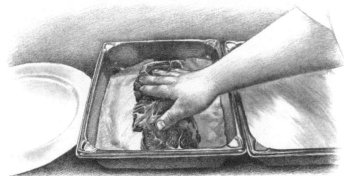

The meat is pressed into the Basic Dry Cure (page 39).

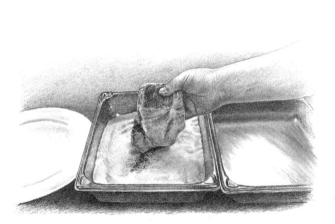

The meat is flipped and the other side and edges are pressed into the cure, packing it on.

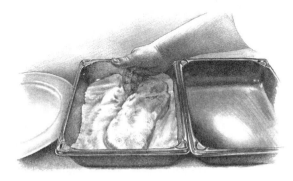

Some cure can be scooped up and pressed onto the top to ensure that the entire surface is well coated.

The meat is put in a pan and left to allow the salt to do its work. This method will cure most meats that aren't more than a few inches thick. The salt is rinsed off the meat before finishing it. Here, for Tasso ham, a heavy aromatic seasoning mixture will be used to coat the slice of shoulder before it goes into the smoker.

5 pounds/2.25 kilograms boneless pork shoulder butt, sliced
crosswise into 5 even slabs

Basic Dry Cure (page 39) as needed for dredging
(about 2 cups/450 grams)

3 tablespoons/30 grams ground white pepper

1½ tablespoons/15 grams cayenne pepper

3 tablespoons/6 grams dried marjoram

3 tablespoons/24 grams ground allspice

1. Dredge the pork in the dry cure, pressing it into the cure to make it adhere, and shake off the excess. The surface of each piece should be coated with an even layer of cure. Refrigerate, covered, for 4 hours.

2. Rinse the pork under cold water, brushing off any remaining dry cure, and pat dry with paper towels.

3. Combine the remaining ingredients in a shallow bowl, mixing well. Dredge the meat on all sides in the spices so that the pieces are uniformly coated.

4. Hot-smoke the pork (see page 72) to an internal temperature of 150 degrees F./65 degrees C. The ham will keep for several weeks refrigerated and for several months well wrapped in the freezer.

Yield: Five 14-ounce/350-gram slabs tasso ham

Smoked Pork Loin

Here are two recipes for pork loin, both of which take into account that this particular cut of pork, given current commercial growing conditions, is very lean and difficult to cook flavorfully.

CANADIAN BACON (CURED SMOKED PORK LOIN)

Canadian bacon is brined with a curing salt and herbs before it's smoked. The brine, of course, results in a moister loin than one that has not been brined. The herbs season it, the pink salt gives it its cured flavor, and then the smoke adds a third layer of flavor.

Alternatively, you can make perfectly tasty Canadian bacon without smoking it, roasting the loin in a low oven (200 degrees F./93 degrees C.) to an internal temperature of 145 to 150 degrees F./63 to 65 degrees C. But the smoke really does add an important dimension to this lean cut.

THE BRINE

1 gallon/4 liters water

1½ cups/350 grams kosher salt

1 cup/225 grams sugar

6 teaspoons/42 grams pink salt

1 large bunch fresh sage

1 bunch fresh thyme

2 garlic cloves, peeled and lightly smashed

One 4-pound/2-kilogram boneless pork loin, all fat
and sinew removed

1. Combine all the brine ingredients in a pot large enough to hold the pork loin and bring to a simmer, stirring to dissolve the salt and sugar. Remove from heat and let cool to room temperature, then refrigerate until chilled.

2. Place the pork loin in the brine and weight it down with a plate to keep it completely submerged. Refrigerate for 48 hours.

3. Remove the loin from the brine (discard the brine), rinse it under cold water and pat dry. Place it on a rack set over a plate or a tray and refrigerate, uncovered for 12 to 24 hours.

4. Hot-smoke the pork (see page 74) to an internal temperature of 150 degrees F./65 degrees C., 2 to 3 hours. Allow to cool, then cover and refrigerate for up to 10 days.

Yield: 3½ pounds/1.5 kilograms Canadian bacon

SPICY SMOKE-ROASTED PORK LOIN

In this recipe, a complex, highly spiced dry rub makes a plain pork loin extraordinary. The smoke does add a superb depth but the dry rub is so intense, the pork can also simply be roasted with excellent results (in a 350-degree-F./175-degree-C. oven to an internal temperature of 145 degrees F./63 degrees C., about 40 minutes). To slow-roast it on a rotisserie would be another outstanding alternative, or grill it over low or indirect heat to that same internal temperature.

The loin can be served immediately or refrigerated and eaten cold, sliced thin. Hot or cold, this spicy pork will go well with a sauce using sharp flavors such as mustard or horseradish. If you want to reheat the cold pork, do so in a moist environment, on a bed of sliced onions or cabbage in a pot with some stock and aromatic seasonings, to keep it from drying out.

One 4-pound/2-kilogram boneless pork loin
Spicy Dry Rub (recipe follows)

1. Coat the pork all over with the dry rub. Wrap it in plastic and refrigerate it for 12 to 48 hours.

2. Smoke-roast the pork (see page 72) at 325 degrees F./160 degrees C. to an internal temperature of 140 degrees F./60 degrees C., about 45 minutes.

Yield: 3½ pounds/1.5 kilograms smoked pork loin; 8 to 10 servings

Spicy Dry Rub for Pork

This dry rub is excellent on just about any cut of pork, shoulder, loin, or tenderloin, whether you plan to smoke it, roast it, or grill it. Because it's so strong, you won't need a lot of extra sauce if you're using it with shredded pork shoulder. When using it with large cuts, it's best to apply the rub to the meat ahead of time, then wrap it in plastic and refrigerate for 1 to 3 days.

1 tablespoon/10 grams freshly ground black pepper
1 teaspoon/3 grams cayenne pepper
2 tablespoons/16 grams chile powder
2 tablespoons/16 grams ground coriander
1 tablespoon/13 grams dark brown sugar
1 tablespoon/2 grams dried oregano
4 tablespoons/32 grams Spanish paprika
3 tablespoons/40 grams kosher salt
1 tablespoon/10 grams ground white pepper

Combine all ingredients, mixing well. Store in an airtight container, the rub will keep well for up to 4 months.

Yield: About 3/4 cup/150 grams

PASTRAMI

Pastrami differs from corned beef in two main ways: it's smoked and it's coated with a combination of crushed coriander seeds and black peppercorns. Other than that, it's corned beef underneath a smoky crust. The coriander-peppercorn crust is a spicy flavorful seasoning that works well on most meats, as well as fish and poultry.

The best way to serve pastrami is to reheat it gently for hours until it's falling-apart tender (steamed or gently roasted, see step 7 below). We favor the beef plate, a cut from the breast or below the shoulder: because it is especially fatty, it stays juicy after all the cooking, and thus makes an especially fine sandwich. Any good butcher shop will be able to provide you with this cut.

THE BRINE

1 gallon/4 liters water

1½ cups/350 grams kosher salt

1 cup/225 grams sugar

6 teaspoons/42 grams pink salt

1 tablespoon/8 grams Pickling Spice (page 68)

½ packed cup/90 grams dark brown sugar

¼ cup/60 milliliters honey

5 garlic cloves, minced

One 5-pound/2.25-kilogram beef plate or beef brisket,
        heavy surface fat removed

1 tablespoon/8 grams coriander seeds, lightly toasted (see
        Note page 51)

1 tablespoon/10 grams black peppercorns, lightly toasted (see
        Note page 51)

1. Combine all the brine ingredients in a pot large enough to hold the beef and bring to a simmer, stirring until the salt and sugar are dissolved. Remove from the heat and allow to cool to room temperature, then refrigerate until the brine is chilled.

2. Place the beef in the brine and put a plate on top of it to keep it completely submerged. Refrigerate for 3 days.

3. Remove the beef from the brine, rinse it, and dry it. Discard the brine.

4. Combine the coriander and pepper in a spice mill or coffee grinder and pulse until coarsely ground. Coat the beef evenly with the mixture.

5. Hot-smoke the beef (see page 74) to an internal temperature of 150 degrees F./65 degrees C. (Traditionally, pastrami is first cold-smoked, then hot-smoked to achieve a heavy smoke. So try to get as much smoke on it as possible by keeping it below its final temperature for as long as possible.)

6. To prepare the pastrami for serving, preheat the oven to 275 degrees F./140 degrees C.

7. Place the beef in an inch of water in a stockpot or on a rack above the same amount of water in a roasting pan. Bring the water to a simmer, then cover the pot, place it in the oven, and slow-roast or steam for 2 to 3 hours, until it's fork-tender.

Yield: 5 pounds/2.25 kilograms pastrami

CAROLINA-STYLE SMOKED BARBECUE

You could call this pulled pork, but as anyone from the Carolinas will tell you, *barbecue* is the proper term for shredded pork mixed with a tangy vinegar-based sauce and served, on a bun or by itself with a pile of hushpuppies, and a big glass of iced tea. Here a bone-in pork shoulder butt is seasoned and smoked for several hours. If you'd like additional seasoning, use the Spicy Dry Rub for Pork on page 88 in place of the salt and pepper. The recipe can also be prepared by grilling or roasting the smoked pork (covered, at 250 degrees F./120 degrees C. for 4 to 6 hours, until meltingly tender), or a combination of the two. No matter how it's cooked, the resulting meat shreds easily.

Once the meat is shredded, a tangy sauce is stirred in; the amount of vinegar in a proper Carolina barbecue sauce depends upon your longitude as does the seasoning. Western Carolina sauces include tomato in some form, whereas eastern ones do not, though the latter often contain mustard. No matter where it comes from, it's hotly debated and fiercely contested, and those fond of regional distinctions can get really worked up about the subject. We find it better to eat the subject.

1 bone-in pork shoulder butt, about 5 pounds/2.25 kilograms
3 tablespoons/40 grams kosher salt
2 teaspoons/7 grams freshly ground black pepper
Carolina-Style Barbecue Sauce (page 290)

1. Sprinkle the pork evenly with the salt and then the pepper.

2. Hot-smoke the pork (see page 74) at between 200 and 300 degrees F./93 to 150 degrees C. for 3 hours.

3. Transfer the pork to a braising vessel with a lid, such as a Dutch oven or a roasting pan covered with foil, and add ½ cup/125 milliliters water. Cover tightly and set in the oven. Turn the oven to a 250 degrees F./120 degrees C. and cook until the meat shreds easily, about 4 hours.

4. Shred the meat in the pot. Add the barbecue sauce, stirring till the meat is evenly coated with the sauce. Serve warm.

Yield: 4 pounds/2 kilograms pulled pork

AMERICAN-STYLE BROWN-SUGAR–GLAZED HOLIDAY HAM

This is the classic American baked ham, like the honey-baked ham most are familiar with—as worthy a ham tradition as those of Europe. The ham is brined, then hot-smoked and glazed, after which it can be eaten cold or baked in a low oven to rewarm.

THE BRINE
1 gallon/4 liters water
1½ cups/350 grams kosher salt
2 packed cups/360 grams dark brown sugar
6 teaspoons/42 grams pink salt

One 12- to 15-pound/5.5- to 6.75-kilogram fresh ham, skin and
    aitch-bone removed

THE GLAZE

1½ packed cups/270 grams dark brown sugar

¾ cup/185 milliliters Dijon mustard

1 tablespoon/20 grams minced garlic

1. Combine all the brine ingredients in a container large enough to hold the ham and stir to dissolve the salt and sugar. Submerge the ham in the brine, weight it down to keep it completely submerged and soak for 6 to 8 days (half a day per pound/450 grams).

2. Remove the ham, rinse it under cool water, and pat dry. Place it on a rack set on a baking sheet and refrigerate it, uncovered, for 12 to 24 hours.

3. Hot-smoke the ham (see page 74) at 200 degrees F./93 degrees C. for 2 hours.

4. Meanwhile, mix the brown sugar, Dijon, and garlic in a bowl until smooth. Brush the ham with glaze (reserve the remainder) return it to the smoker, and smoke until it reaches an internal temperature of 155 degrees F./68 degrees C.

5. Remove the ham from the smoker and brush with the remaining glaze. Allow to cool, and refrigerate.

6. To serve, slice and serve cold, or reheat it in a 275-degree-F./140-degree-C. oven until warm in the center (test with a metal skewer).

Yield: 16 to 18 servings

SMOKED SALMON

This is cold-smoked salmon, which means it's first cured (see illustrations on page 52) like gravlax but has the additional flavors of smoke and sweet spices. It's important that the temperature in the smoker never become hot enough to cook the fish, or it will change the texture and flavor of the salmon completely—you'll have hot-smoked rather than smoked salmon, big difference. The salmon is particularly fine when smoked over cherry or another fruit wood, such as apple or pear. Once it's smoked and chilled, slice it thin and serve it as you wish, with bagels and cream cheese, or on toast points with the traditional garnishes of chopped hard-boiled egg, diced red onion, capers, and crème fraîche. Or mince it to make smoked salmon tartare.

THE DRY CURE

½ cup/125 grams kosher salt

¼ cup/50 grams dark brown sugar

¼ cup/100 grams sugar

1 teaspoon/7 grams pink salt

1 teaspoon/3 grams ground white pepper

1 teaspoon/3 grams ground allspice

1 teaspoon/3 grams ground bay leaf

½ teaspoon/2 grams ground cloves

½ teaspoon/2 grams ground mace

1½ pounds/675 grams salmon fillet in one piece, skin on,
　　　pinbones removed

1½ tablespoons/22 milliliters dark rum

1. Mix all the dry cure ingredients together. Choose a dish or pan in which the salmon will fit snugly; it will release copious liquid and form a brine that should remain in contact with the fish. Spread half of the dry cure in an even layer the width and length of the salmon in your container, then place salmon skin side down on the cure. Sprinkle the rum over the fillet, then coat the fish with the rest of the dry cure, coating the thick parts heavily with the mixture and the tapered belly with less.

2. Cover the salmon with plastic wrap, set another pan on top, and place about 8 pounds/4 kilograms of weights on it to help extract the moisture (cans or a few bricks will suffice). Refrigerate for 36 hours, or until the thickest part of the salmon feels dense and stiff to the touch.

3. Remove the salmon from the cure and rinse thoroughly under cool water. Pat dry. Place on a rack set over a tray, and refrigerate, uncovered, for 4 to 24 hours.

4. Cold-smoke the salmon (see page 74) at 54 degrees F./12 degrees C. for about 6 hours, or to taste. Remember that the temperature of the smoker and the salmon should never go above 90 degrees F./32 degrees C.

5. Store the salmon wrapped in butcher's or parchment paper in the refrigerator. It will keep for up to 3 weeks (change the paper if it becomes too moist).

Yield: 1 pound/450 grams smoked salmon

SMOKED SCALLOPS

Sea scallops will take on a range of flavors, depending what they're smoked with, but fruitwoods are best, for their delicate flavor and the pale tan color they impart. Once the scallops are cold-smoked (it's important that temperatures don't go high enough to cook them, which would begin to happen at around 115 degrees F./45 degrees C. or so) they're ready to grill, sauté, roast, or broil. They can be served hot, as a main course, but they are also tasty cold, perhaps as the centerpiece of a composed salad.

If you have bay scallops instead of sea, reduce the brine time to about twenty minutes. Peeled and deveined shrimp are also excellent brined (about half an hour for medium to large shrimp) and smoked this way.

THE BRINE
½ recipe Basic Fish Brine (page 59), replacing 1 cup/
    250 milliliters of the water with white wine
1 medium onion, sliced
1 bunch fresh thyme
8 black peppercorns

3 pounds/1.3 kilograms sea scallops, tough side muscles
    removed

1. Combine all the brine ingredients in a large pot and bring to a boil. Remove from the heat and let cool, then refrigerate until completely chilled.

2. Put the scallops in the brine and weight them down with a plate to keep them completely submerged. Refrigerate for 1 hour.

3. Remove the scallops from the brine (discard it), rinse them, and pat dry with paper towels. Place them on a smoking rack and cold-smoke (see page 74) for 1 to 1½ hours, or until they take on a rich golden color.

4. Remove from smoker and store in the refrigerator until ready to cook. They'll be good for up to 5 days.

Yield: 3 pounds/1.3 kilograms smoked scallops

SMOKED JALAPEÑOS

These are very close in flavor to the chipotles you find dried or canned in adobo sauce. Not surprising, since chipotles are in fact smoked jalapeños. This recipe delivers the characteristic smoke and heat, but with the flavor of a fresh pepper. The method is simple and can be applied to any fresh chile pepper: halve them, cold-smoke them, broil them to char, and remove the skin—and they're ready to go. Fresh smoked chiles are a lot of fun because they add a kind of beguiling sweet-hot-smoky flavor people can't quite place. Use them as a seasoning in salsas, in a spicy rice pilaf, in warm or cold salads, in bean salads, or in a spicy vinaigrette.

12 jalapeño chiles, halved lengthwise

1. Cold-smoke the peppers (see page 74) skin side down on a rack for 3 to 5 hours.
2. Preheat the broiler. Remove the chiles from the smoker, place them skin side up in a baking sheet, and broil to char the skin, about 10 minutes or so, depending on your broiler (if you've hot-smoked or grilled them, this step to remove the skin may not be necessary). When the skin is charred, place the chiles in a bowl, cover with plastic wrap and let cool.
3. When cool, stem, peel, and seed the peppers. Refrigerate covered for up to 6 days, or freeze until ready to use.

Yield: 12 smoked jalapeños

SPICY SMOKED ALMONDS

Almonds take smoke beautifully and, if you further season them with the spicy mixture in this recipe, are very difficult to stop eating. In fact, these are delicious even if you omit the smoking step and simply toss them with the oil and spices and roast them. Alternatively, you could grill these in a steamer basket, metal colander, or in aluminum foil over charcoal. The spices can burn easily, so be careful not to let them go too long in the oven. Combine the salt and spices in a coffee or spice mill and grind so that the salt is fine enough to adhere to the nuts.

1 pound/450 grams unblanched raw almonds

2 tablespoons/30 milliliters vegetable oil

¼ teaspoon/0.5 gram cayenne pepper

½ teaspoon/1 gram garlic powder

1 tablespoon/15 grams finely ground kosher salt or
      fine sea salt, or more to taste

2 teaspoons/6 grams chile powder

1 teaspoon/3 grams freshly ground black pepper

1. Lay the almonds on a rack, such as a mesh cooling rack: if the slats are too wide, cover the rack with foil and poke a lot of holes in it (or use a metal colander instead).

2. Cold-smoke the almonds (see page 74) for 2 to 3 hours.

3. Preheat the oven to 375 degrees F./190 degrees C.

4. Combine the oil and spices in a bowl and toss with the almonds to coat. Spread them on a baking sheet and roast for 10 to 15 minutes, stirring several times. Remove from oven, taste, and toss with more salt if needed.

5. The nuts can be stored in an airtight container for up to 2 weeks.

Yield: About 4 cups/450 grams

4.

✦❖✦

THE POWER AND THE GLORY:
ANIMAL FAT, SALT, AND THE PIG COME TOGETHER IN ONE
OF THE OLDEST, DIVINE-YET-HUMBLE
CULINARY CREATIONS KNOWN TO HUMANKIND

. . . .

Sausage is the apotheosis of the techniques in this book. It involves craftsmanship in the kitchen, care from the cook, and devotion from the eater. There may be no finer package of protein, fat, and seasonings than that which resides within the transparent but resilient hog casing—and none more humble. In this chapter, we discuss all matters relating to all manner of sausages and describe how the sausage is perfected, what turns a good sausage into a great one.

Embrace the sausage. It's an extraordinary luxury available to everyone, a perfect package of seasonings and juiciness unequaled by any category of prepared food, and we wish more home cooks would take advantage of it. An acquaintance of ours who was at the time deputy director of food services in Washington for the House of Representatives was asked to prepare a huge meal in honor of Julia Child's seventieth birthday. Not wanting to disappoint America's most important culinary icon, he created a menu designed to woo any Francophile: oeufs à la Chimay, pantin aux épinards, pissaladière, and other classic dishes. After reviewing his menu, Julia responded: "I really like sausages." John quickly revised the menu to include Cajun andouille, boudin blanc, Toulouse saucisson, and truffle-laced pork and veal sausage in brioche. The menu was met with approval. Julia knew. Sausage is great to eat. Period.

As sausage advocates, Brian and I are also defenders of the good name of fat. When sausages are delicious, it's usually because they're loaded with it. Too much fat tends to be the main criticism leveled at sausage, but saying you'd love sausage if only it didn't have so much fat is like saying you'd love water if only it weren't so wet. Fat is fundamental to the quality of a sausage, its succulence and flavor. If you must avoid fat for dietary reasons, avoid sausage; we don't recommend trying to make or work with low-fat sausages–low-fat sausage is an oxymoron to us.

Store-bought sausages can be good, and some are very good. But rarely are they as satisfying as those you can make at home, because you can adapt homemade sausage to your own tastes. You can use the best cuts of meat as opposed to the butcher's scraps. If you like spicy heat, you can add more, or vary the type of heat–ordinary red pepper flakes can be replaced with ground chipotle pepper or Espelette powder, the superb cayenne-like pepper from the Spanish town of the same name. Or perhaps you like sweet spices such as nutmeg and clove for seasoning sausage. If you don't care for pork, turkey and chicken make superlative sausages without compromise. One of the tastiest, juiciest sausage recipes in this book is a chicken sausage with basil, tomatoes, and garlic.

If you do like the flavor of pork, sausages are another of the glories offered by the pig–thus sausage's place as the linchpin of the charcutier's trade. While all kinds of meat have been turned into sausage, pork has dominated sausage making since antiquity because of the flavor, quality, and quantity of both its meat and fat. Sausage is also a key garde-manger technique. Today the *garde manger* is responsible for cold food in a restaurant kitchen–salads and the like–but traditionally this member of the *brigade de cuisine* was an ingenious scavenger, turning leftovers and trimmings into finished dishes. And the sausage's original raison d'être was to make use of the trimmings after the main cuts of the pig had been broken down.

For a chef, sausage making is also an effective way to stretch food. Not long ago, for instance, Brian was intending to serve venison loin for a special game dinner at the restaurant and so had ordered a saddle of venison. At the last minute, the number of guests increased, and once he'd removed the loins from the saddle, he realized that they weren't big enough to feed all the guests. He reconsidered his plan: as he had the whole saddle, and therefore abundant scraps, he decided to season those scraps with paprika, nutmeg, allspice, garlic, and pepper to transform it into a tasty venison sausage (see page 155 for the recipe) rather than to buy another expensive loin at the last minute. The sausage allowed him to serve smaller portions of the venison loin, without reducing the value of each plate. Indeed, he increased the value: he'd not only solved his dilemma, he'd also introduced an additional component to elevate the entrée.

There's no reason the same thinking can't work at home, in terms of what chefs call "total utilization"—or most of us would refer to as "using everything"—extending not-enough-food into exactly enough. Furthermore, knowing how to manipulate food gives you more freedom in the kitchen. For example, if you had to have dinner ready in an hour, and all you had in the fridge was a tough cut of meat, say a lamb shoulder, something that would require hours of slow braising to become tender, you could instead grind it, season the meat with salt and pepper, maybe a little garlic, maybe some herbs (parsley, rosemary, thyme—whatever you happened to have on hand), maybe some chopped sautéed mushroom or onion for moisture, or fat (pork or lamb). You could then sauté this impromptu sausage, as is or formed into patties, or you could wrap those patties in cabbage or grape leaves and braise them. Served with a couscous or pilaf and some vegetables, they'd make an exotic meal. By turning the lamb shoulder into sausage, you are merely changing the tenderizing mechanism from an atmospheric one (hot, moist heat) to a mechanical one (grinding), but with dramatic differences.

Moreover, while a sausage is prized for its flavor, and is rightfully considered a special item, it requires the least expensive cuts, tough cuts with a lot of connective tissue and plenty of good fat marbled in. Making sausage is a terrifically satisfying way to serve a budget cut of meat.

Food lovers tend to think of sausage as ground meat in a casing, but lots of traditional sausages don't use a casing. So not having a stuffer or sausage casing is no reason to dismiss sausage recipes. Loukanika, the Greek sausage flavored with spices, herbs, and orange peel, is formed into patties that are grilled, roasted, or sautéed—like American breakfast sausage. Often the loukanika patties are wrapped in caul fat, a fatty membrane from the stomach of pigs and lambs (sausages wrapped in a casing of caul fat—see page 103—are sometimes called *crèpinettes*, from the French word for caul, *crèpine*). Mexican chorizo is another freeform sausage, used as an ingredient in various dishes, and Italian sausage is often used loose, its casing removed. Even the more complex of what are technically called emulsified sausages, sausages made of meat and fat processed to a finely textured paste, don't always require casings. A good bratwurst mixture can be squeezed out of a piping bag straight into hot water and poached, then chilled and kept in the fridge until it's ready to fry (*bratwurst* means frying sausage, from *braten*, to fry). The same technique is often used for religious reasons with veal or beef sausages, for those who want to avoid pork casings.

Sausage usually becomes more special, however, when it's stuffed into casings, and this type of sausage making can seem like a project in the home kitchen. Once you set yourself up to stuff sausage into casings, though, the capacity for your pleasure and for pleasing others is increased tenfold. They're fun to make, and the casing itself is a superlative natural cooking vessel. You do need the right equipment and the right ingredients, and you do need to plan ahead. The technique is a multistep process with plenty of cleanup throughout. Clear a good two- to three-hour window for leisurely kitchen work when making sausage.

Even more important than making any sausage properly, however, is cooking it properly. You can make a beautiful sausage but it will be disappointing if you don't cook it properly. Sausage may be the most commonly overcooked protein in America, after the chicken breast. It is typically either cooked without thought or cooked with fear of not cooking it enough. A sausage with a split casing, all its juices running out, is a badly cooked sausage. A dry sausage is an overcooked sausage.

How to cook a sausage? Cook it the way you would a pricy beef tenderloin or rack of lamb: carefully and to a precise internal temperature.

Most meat sausages have some sort of pork in them (if only just fat). Standard recommendations for cooking pork are that it be cooked to a temperature of 150 degrees F./65 degrees C., for a final temperature of 155 degrees F./68 degrees C. We personally never want to cook a pork chop or loin that far, but for a pork sausage, it's fine, even prudent, and the sausage remains juicy at that temperature because of its high fat content and because the flavor and juiciness are contained in the casing. For safety reasons, chicken and turkey sausages must be cooked to ten degrees higher, for a final temperature of 165 degrees F./73 degrees C. Unfortunately, most people seem to cook sausages past 200 degrees F./93 degrees C., until they're grainy and dry inside. Indeed, when you've eaten a truly great sausage, its greatness may well have been the result not of the meat or the seasonings but of proper cooking.

So sauté, roast, grill, or poach your sausage and check it using an instant-read thermometer. Remove the sausage when the internal temperature reads five degrees lower than you want it to be; carryover cooking after it's removed from the heat will take care of the rest. When you do that, you'll know how truly magnificent a perfectly cooked sausage is.

This chapter divides sausages into three categories—fresh, emulsified, and smoked—though many sausages actually fall into more than one category. All the recipes, because they are based on ratios, can be successfully halved or doubled according to your needs. Brian and I decided to develop these recipes mainly for five-pound quantities because making sausage is a project and because sausage freezes well, it's wise to make plenty; but five pounds is the maximum amount most home equipment can handle in a single batch. Each recipe yields about twenty 6-inch/15-centimeter links if using standard hog casings.

FOR COOKS WHO WANT TO MAKE SAUSAGE BUT DON'T EAT PORK

We're well aware that many of our readers cannot eat pork. Others simply, in the words of Samuel Jackson in *Pulp Fiction*, don't "dig on swine." That doesn't mean that sausage making is unavailable to you. While the fat issue does present an obstacle (pork fat is what makes most of these sausages succulent), there's no reason you can't use other fats, whether chicken or olive oil. We've even made a chicken sausage with butter with excellent results.

The general rule though is simply to choose the fattiest cuts of whatever type of meat you're using. If you're working with leaner cuts, you can add beef suet instead of pork fat for additional succulence. Of course, all other sausage-making rules apply, regardless of the type of meat and fat you use.

About Casings

What would seem to be a trash part of the animal–the inner lining of the intestine–is actually an extraordinary cooking vessel. It's very fine and light but very strong. It is a receptacle for the meat and juices and yet when the sausage is hung to dry, the casing is porous enough to allow moisture to escape so that we can have dry-cured *salume, saucissons sec,* arguably the pinnacle of the sausage-maker's art. Casings brown beautifully, adding flavor and color to sautéed or grilled sausages. And there's nothing like the bite a casing gives when you're eating a sausage from a bun–that snap, before the plunge into fatty juiciness. If you took away that snap, the fatty juiciness would still be good, but that pop, that almost crunch before you get there, is a pleasure unique to sausage eating.

Hog casings, which come from the small intestine, are easy to find, and, thanks to the salt in which they're usually dry-packed, will keep for a year in your refrigerator. Casings packed in brine will last about a month refrigerated. The meat department at your supermarket should be happy to order casings for you (and most other hard-to-find products, for that matter, such as fresh fatback). Specialty markets and butchers who make their own sausage are often willing to sell you casings over the counter. And numerous companies now offer casings by mail (see Sources, page 303).

Hog casings, which are about an inch and a half in diameter when stuffed, are used for most of the sausages in this book, from fresh brats to Italian to smoked kielbasa. Most other types of casings need to be ordered from sausage supply companies: sheep casings, which are smaller and more delicate, used for breakfast links, chipolatas, hot dogs, and smoked andouille; larger hog casings, which include hog "middles" and, from the lower part of the intestine, hog bungs, the very end of the digestive tract (this is what's used for the famous rosette de Lyon); and beef casings, the big boys–rounds for blood sausage and ring bologna, bungs for salamis and large bologna, and the bladder for traditional mortadella.

All casings should be soaked for at least 10 to 20 minutes before using, or for as long as 2 days. This will remove the salt and any residual odor as well as rehydrate them. If you soak for a short time, do so in tepid to warm water, to hasten the process. If you are using bungs, from the lower part of the intestine, or other larger casings that have an odor, soak them under a thin stream of running water or in frequent changes of water until all of the odor is gone. Once the casings are rehydrated, hold them open under running water to flush out the insides. All casings should be completely clean and odorless before you use them.

Caul fat, the veil-like connective membrane that surrounds the stomach and other viscera of sheep and pigs, called the omentum, can also be used as a kind of casing. Essentially a layer of collagen and fat that melts away during cooking, caul fat is an extraordinarily useful kitchen tool. Roasts can be wrapped in it for extra moisture. Braised items can be wrapped in it and reheated. And sausage patties can be wrapped in it and roasted. The caul fat serves the same function as the casing, retaining moisture, maintaining shape, and browning for flavor and color.

Also available are natural processed collagen casings, animal collagen tissue extruded into the shape of sausage casing. The material looks a little like thin, tough wax paper and despite its unnatural feel, it works well for breakfast sausages. Other larger collagen casings are available too. There are even plastic casings, colored casings, and casings lined with spices that, when the casing is removed, remain on the sausage, but generally there's no reason to stray from natural casings.

The recipes in this chapter most often call for the common hog casing; several recommend sheep casing and a few, hog bung or beef casings. But any of the sausage recipes can be stuffed into whatever kind of casing you have–don't let a casing prevent you from trying a recipe that intrigues you. You will need about two feet of hog casing or four feet of sheep casing for each pound of sausage, but it's a good idea to soak a few extra feet to ensure you don't run out mid-stuff.

If you buy hogs directly from a grower and want to process your own casings, work with five-foot lengths of the small intestine that have been flushed at the slaughterhouse. While they are still warm, flush them again with plenty of water. Next, turn them inside out. Begin by folding over one end, as you would when turning a sock inside out. Hold the fold beneath running water; the weight of the water will turn the casing inside out for you. The hard part: On a cutting board, drag a knife over the intestine to remove the thick mucous lining of the intestine until you are left with only the white transparent membrane; it takes a good deal of scraping. Store the intestine packed in plenty of kosher salt and refrigerated. Use a strong bleach solution to sterilize your board, sink, counter, and knives. It's an education and valuable to do for that reason—I did it once, and that was enough; it's laborious and generally pretty disgusting, even for people who enjoy on-the-farm cooking. We recommend purchasing even if you have plenty of warm innards available to you.

Other Special Ingredients for Making Sausage

- **PORK SHOULDER BUTT**: The cut from the shoulder of the pig, above the front leg (which is called the picnic ham); also called pork shoulder, shoulder butt, or Boston butt. Pork butt is heavily marbled and inexpensive, and thus the perfect cut for sausage making.

- **BACK FAT (OR FATBACK)**: The layer of fat from the pig's back, usually the purest white and thickest fat on the pig's body. For use in sausages and pâtés, this can be special-ordered by your butcher or grocery store or ordered online; make sure to request fresh back fat, not salted back fat. But jowl fat is actually considered best by charcutiers because it's especially creamy. And some people who make their own lard will search far and wide for the fat that surrounds the pig's kidney, for its delicate flavor and texture.

- **PINK SALT**: A curing salt containing sodium nitrite, used for sausages and other meats that will be smoked. It goes by many names depending on the company selling it (Prague Powder #1 or Insta Cure #1, DQ Curing Salt, tinted cure mix or TCM), but we refer to it here generically as pink salt (see Sources, page 303). Regardless of what it's called, it is salt with 6.25 percent sodium nitrite added, and it is tinted pink to prevent accidental ingestion. If it's ingested in any quantity, it can be harmful, even lethal. See pages 174–176 for important details and warnings about nitrites, and keep this candy-colored salt out of the reach of children.

- **DQ CURING SALT #2 AND INSTA CURE #2:** This pink salt with nitrate (see Sources, page 303) is used only for sausages that are dried for many days or weeks, saucisson sec or salami. Nitrate converts to nitrite over time, acting like a time-release cold capsule, to prevent the growth of the bacterium that causes botulism. The same warnings that apply to pink salt apply to this curing salt as well (see pages 174–176 for details).
- **DEXTROSE:** A type of refined corn sugar. Its flavor is not overly sweet and, in fermented sausages, it's the perfect food for lactic bacteria to feed on (for more on this, see page 176).
- **STARTER CULTURES:** Starter cultures, such as Bactoferm F-RM-52 (see Sources, page 303), are a live bacteria that feeds on sugar and produces acid, sometimes added to dry-cured sausages to lower the pH by generating acid. This reaction results in an environment in the sausage that prevents the growth of unwanted microorganisms and gives the finished sausage a delicious tanginess found in Old World–style dried sausages.
- **FERMENTO:** A dairy-based flavoring, used to imitate a fermented flavor (see Sources, page 303).

Fresh Sausage

Fresh sausage means meat ground with seasonings, cooked, and eaten hot. It's not much more complicated than that, but even with fresh sausage, a few issues of technique must be respected in order to achieve a superlative sausage.

Basic Fresh Sausage Technique

SPECIAL TOOLS

Having the right tools makes the work easier. These help considerably:

Scale

5–6 quart meat grinder (a KitchenAid meat grinder
attachment is adequate for occasional use)

KitchenAid mixer with paddle

Sausage stuffer

Food processor

Sausage casings

Instant-read thermometer

Brian uses a professional-grade meat grinder. Michael uses a Weston grinder for home countertop use.

Use fresh ingredients. That applies to dried herbs too when a recipe calls for them: the marjoram that's in the spice rack you got as a wedding gift ten years ago will taste like what it is, old dried leaves.

As a rule, it's best to cut your meat and fat into cubes (a dice small enough to fall through the feed tube of your grinder–you shouldn't need to mash the pieces through), removing sinew as you do. Combine the meat and fat with your seasonings a few hours, or as much as a day, before grinding. Seasoning any meat in advance almost always improves it, and salting early also encourages a uniform distribution of the seasonings. Salting pork ahead of time is especially effective, no matter what cut or what you're using it for.

Kosher salt is the most important seasoning in your kitchen, and in any sausage. As a rule, Brian and I use 1.75 percent of the weight of the meat and fat. Often weights aren't exactly 5 pounds or 2.25 kilograms. If we have say, 70 ounces of meat or 1.3 kilograms, we multiply the weight by .0175 (1.75 percent) to determine how much salt to add. Most of the recipes in this chapter are for 5 pounds of meat, requiring 1½ ounces/40 grams of salt (about 3 tablespoons Morton's kosher salt, see page 38), a smidge over 1.75 percent. But salt tolerances differ. Start with this ratio, and add or scale back according to your tastes. (Remember, too, that different brands of salt have different densities, so be careful if you're measuring salt by volume–that is, with a measuring spoon–rather than by weight; see page 28.)

After salt, other seasonings are a matter of taste. Just about anything goes, which is another reason sausage making is fun. If you like an aromatic sausage, add a generous amount of fresh herbs. Cooked diced vegetables, such as onion or roasted red pepper, can be another excellent addition. If you like heat and spice, add dried ground chile peppers and ground toasted coriander seed.

If you're inventing your own sausage, stick to pairings you know work no matter what the form. You wouldn't season a leg of lamb with fresh dill; likewise, you wouldn't make a lamb and dill sausage. But rosemary and garlic go great with leg of lamb; they'll do the same with a lamb sausage. Want to make an Asian-flavored sausage? Season it with garlic, scallions, and ginger. You could stuff that mixture into a casing for a sausage or wrap dough around it for a potsticker. You like the flavors associated with the Southwest? Try adding minced chipotle chiles and lime zest and, just before mixing, kernels of fresh corn. In fresh sausage making, once you understand the ratios, you're limited only by your imagination.

KEEPING YOUR MEAT COLD

It's very important to keep your meat as cold as possible during the sausage-making process. Sausage that gets too warm can "break," meaning the fat and the protein will separate from each other when cooked. You can't always see this happen when you're making the sausage, but it results in a dry, crumbly texture, rather than a smooth, firm, juicy bite. Two tips: Don't leave your meat out at room temperature while you ready your ingredients, and always grind the meat and fat into a bowl set in ice, or partially freeze your meat and fat before grinding.

There are two scenarios that work best. Cut up your meat and fat, add the salt and additional seasonings, and refrigerate the mixture until it's thoroughly chilled, for several hours, or overnight. Or, cut and season the meat and place it in the freezer for 30 minutes to an hour, till it's very cold. The meat can be on the brink of freezing, almost crunchy as it were, but not frozen through. Fat can be ground frozen. We recommend chilling your meat to near-freezing temperatures if you're making sausage in a hot kitchen or a summertime kitchen.

GRINDING

All sausage is by definition ground to some extent (the meat for some country sausages is only finely chopped). Commercial grinders come with a variety of die sizes, but here we'll assume you're using either a standard grinder attachment for a standing mixer or a countertop grinder, both of which come with a large die and a small die, typically $\frac{1}{4}$- and $\frac{1}{8}$-inch/0.5- and 0.25-centimeter holes. In the following recipes, use the small die unless otherwise instructed.

Keep your blade and dies clean and dry to maintain sharpness. Wash and dry them by hand and store them securely. A dull blade can ruin the texture of a sausage. It's worth the small expense to have blades and dies professionally sharpened often.

Temperature is important here too: Keep everything cold. This is simply good food safety practice as well as an element of good craftsmanship. If you keep all your ingredients chilled, below 40 degrees F./4 degrees C., usually your meat and fat will combine perfectly. When you grind your meat, do so into a bowl set in ice. It helps to chill the grinder and blades too. All this attention to temperature does make a difference. (If you'd ever ruined 5 pounds of sausage, which ended up with the texture of papier-mâché in your mouth, you'd be vigilant from then on!)

Remove as much sinew as possible when dicing the meat. This can tangle on the blade and clog the die, causing what's called "smear," and this can result in a broken sausage as well. When your grinder is working properly, the meat should be extruded cleanly through the die, each hole in the die distinct, the meat and fat distinct from each other. It should look like good ground beef. If the meat begins to look mushy and, rather than being extruded, collects on the surface of the die in a mass, the color pale because the meat and fat are being squeezed together, turn off the machine, remove the blade, and clear it of sinew. Then resume grinding.

MIXING (THE PRIMARY BIND)

Ground meat, fat, and seasonings need to be vigorously mixed until they are sticky. In charcuterie circles, this is called "the primary bind." Ground meat does not naturally hold together. To make a hamburger that won't fall apart on the grill, you have to work it a little. The more you knead it, the more it sticks to itself and the stickier looking it becomes. The mechanical action of mixing and kneading develops the protein in the meat (as the same action develops the protein in bread dough) and this meat protein, called myosin, is sticky.

Developing the primary bind is an important step in sausage making; it ensures a uniform texture, rather than a coarse, crumbly one, improving both the cooking and the eating, and it ensures even distribution of the seasonings. Mixing can be done with a wooden spoon in a bowl, but if you have a standing mixer with a paddle attachment, that works best.

Mix the sausage for a minute or two to develop that protein, that primary bind, and create a smooth texture. Often during mixing, you will be adding water or wine or some other liquid to the meat, which the meat will absorb quickly and easily. The liquid enhances the moisture of the finished sausage, helps to distribute the seasonings, and, if the liquid is a flavorful one, such as wine, it also functions as a seasoning. This step gives you another opportunity to chill your meat: always add ice-cold liquid.

To mix using a standing mixer: Grind your meat into the mixer bowl, which should be set in ice. Remove the bowl from the ice and attach it to the mixer. Using the paddle attachment, mix the meat and fat for 1 minute on low speed. Add the ice-cold liquid and mix on medium speed for 1 more minute, or until the liquid is incorporated and the mixture feels tacky and has a uniform sticky, almost furry, appearance.

To mix by hand: Grind the meat into a bowl set in ice: it should be big enough to contain the meat during vigorous stirring. Using a wooden spoon, stir and press and fold the meat and fat for about a minute to begin the primary bind. Add the ice-cold liquid and stir vigorously until the liquid is incorporated and the meat mixture coheres and looks sticky. (It takes some muscle.)

THE TEST

It's always a good idea to check your seasoning by making a tiny patty from your mixed sausage and sautéing it (or, in some cases, wrapping it in plastic and poaching it). If there's not enough salt, add more, and be sure to mix long enough to distribute it evenly. If the other seasonings taste weak or out of balance, now is the time to fix them. If your sausage tastes too salty, your only option is to add more meat and/or fat, but sticking to the 1.75 percent salt ratio, we've never found oversalting to be a problem.

STUFFING

No hidden technique here. Soak your casings for 20 minutes or until supple, or for up to two days. Then hold them open beneath cold running water to rinse out the insides. Fill your stuffer and crank or press till the meat appears at the end of the tube. Slide the entire casing onto the nozzle (it helps to use a little of the water out of its soaking dish to lubricate the stuffing tube), and stuff away. Keep your work surface or baking sheet slicked with water too, so the sausage slides. When you're finished, squeeze the sausage into links of the specified size and twist them in alternating directions (or tie off each link with butcher's string, if you wish); this will force any air pockets out toward the casing, and then you can prick these air pockets with a needle, knife tip, or sausage pricker. Cover and refrigerate or freeze your sausages until you're ready to cook or smoke.

Note: If you're using the stuffing attachment of your standing mixer, do not grind right into the stuffer. Grind your meat, and mix it to create the primary bind and incorporate the liquid. Then refrigerate it until it's chilled before sending it back over the auger and into the casing.

COOKING

Sautéing a sausage allows you to develop a tasty, visually appealing skin. Even if you're using the sausage in a stew, start it off with a quick sauté in a little oil to brown it for an appealing color before adding it to the pot. It's important to sauté sausages over medium-low heat so that they cook all the way through without bursting. Film a griddle or a sauté pan, large enough to give your sausages some space between one another, with oil or butter and heat over medium-low heat; heating the pan first will prevent the sausages from sticking. When

SAUSAGE: Seasoning, Grinding, Mixing, Stuffing, Linking

Sausage *mise en place*: meat and fat; ice-cold liquid (typically water or wine); sausage casings, and seasonings, including salt and pepper.

There's more than one way to stuff a sausage casing. If your sausage mixture is pliable enough, it's possible to stuff it into casings using a canvas pastry bag. It's much easier, though, to use a metal stuffer. The above stuffers each have their advantages. The plunger (left) is the less expensive of the two, but it can be messy, with sausage squeezing out around the disk that pushes the stuffing through the tube. The piston-and-crank model is by far the best stuffing device—it's very easy and clean to use and makes stuffing a lot of sausage a breeze—but it's more expensive. If you make a lot of sausage, though, this model may be worth the expense.

The seasonings and liquid are mixed with the meat. The meat can be covered and refrigerated so that the seasonings can fully infuse it, the best scenario, or it can be ground right away.

The KitchenAid standing mixer (right) with a grinder attachment is an adequate tool for grinding sausage. Machines designed solely for grinding (left) are also available and are what we recommend. The feeder tube trays on such models hold plenty of meat, and they typically come with a variety of dies. If the tube section surrounding the auger is metal, though, it can get very hot, so be sure to freeze this piece before you grind. The meat is extruded through the die in distinct cylinders; if the meat bunches up and appears to come out as a blob, a condition called "smear," caused by sinew wrapping around the blade or by dull blades, it can ruin your sausage. If this happens, immediately stop the machine, remove the die, and clean the blade, then resume grinding. It's a good idea to have your blades and dies sharpened professionally once a year or so. Always wash and dry your blades by hand and store them carefully to keep them sharp.

The seasoned meat is passed through the grinder and into a bowl that has been set in ice. Grinding heats the meat, so it's very important to keep the meat as cold as possible; if the meat gets too warm, the fat will separate from the meat when cooked, and the result will be a mealy textured sausage that's unpleasant to eat.

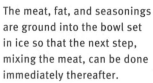

The meat, fat, and seasonings are ground into the bowl set in ice so that the next step, mixing the meat, can be done immediately thereafter.

To ensure that the seasonings are well distributed, that the texture of the finished sausage is tight rather than crumbly, and to incorporate the liquid thoroughly, the sausage mixture is mixed for about two minutes or so with the paddle attachment. When the liquid has been incorporated and the meat takes on a very tacky appearance, it's mixed enough. This step is known as the "primary bind" among sausage aficionados.

The sausage stuffer is filled with the sausage mixture, which is pressed down until the sausage is at the opening of the nozzle, then the casing is fed over the nozzle. Sometimes the casing can stick, so it helps to scoop a little water into the opening of the casing to facilitate feeding it onto the nozzle.

Casings must be soaked in water for 20 minutes or so until supple, or for up to two days, to rehydrate them and remove the salt. They should then be flushed with running water to further clean and hydrate them.

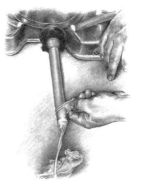

Almost the entire casing is fed onto the nozzle, leaving a couple inches of overhang at the end.

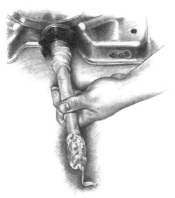

The casing is ready to be stuffed.

It helps to keep a hand on the casing as the sausage is extruded to ensure that the casing is packed firmly but not stretched tight. If air pockets develop, the casing should be pricked with a needle or a knife point. It also helps to slick the counter with water so that the sausage doesn't stick but instead coils neatly.

To ensure a rope of uniform links, it's helpful to measure the links, pinching them at either end to press the sausage gently in either direction to make room for a twist in the casing. Here the sausage is knackwurst (page 151), a stubbier sausage than the customary 6-inch link.

After the the first link is twisted, the second link is measured, but not twisted; then the third link is measured off and twisted in the same direction as the first, which will twist off the second link as well. The process is repeated as necessary. If the sausages will be smoked, the ropes should be cut into even-numbered sections, or there will be one loose sausage hanging from the stick.

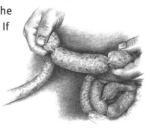

you place the sausages in the pan, you should hear a lively but calm snap and crackle, not a furious one. The sautéing should take some time. As the sausages expand within their casings, they will curl. When the bottom takes on an appealing amber color, turn with a spatula or with tongs, being careful not to puncture the skin, and cover the pan. The sausages will cook a little more quickly and evenly covered, about 10 to 12 minutes. Use an instant-read thermometer to check the internal temperature: 150 degrees F./65 degrees C. for pork and beef, 160 degrees F./71 degrees C. for poultry. Or cut one open to check: Touch the center of the sausage lightly with the back of your finger; it shouldn't burn, but it should be too hot to keep your finger there for more than a moment. It should look cooked, not raw, and when you press down on the sausage, fat and juice should quickly bead out. Slice off a piece and taste it.

Roasting is the easiest way to cook sausages, but you lose the deep caramelization in the casing that you can achieve through sautéing. Still, if you've got a lot going on in the kitchen, this is an effortless way to cook sausages. Preheat your oven to 300 degrees F./150 degrees C. Heat an ovenproof sauté pan with a film of oil over medium-low heat, as you would for sautéing sausage. When the pan is hot, place your sausages in the pan and place the pan in the center of the hot oven until they're done, about 10 minutes. If you have a probe thermometer with a timer, you can be alerted the instant the sausage hits 150 degrees F./65 degrees C. (or 160 degrees F./71 degrees C. for poultry sausage). This little device is not expensive and comes in handy for all kinds of roasting and cooking, and especially for smoking sausages.

Grilling is, of course, the most flavorful way of cooking sausage because sausages carry the flavor of smoke so well. To properly grill sausages, you need to be able to cook them over indirect heat—meaning next to but not directly above the coals—for at least part of the time. If your grill doesn't offer this option, then you need to be able to maintain very low flames. The key to cooking all sausages is gentle heat, uniform ambient soft heat so that the interior cooks before the exterior overcooks or splits open. A few grilling strategies: Start the sausages over medium direct heat, just long enough to give them some color and smoke, then move them to the side, off the heat, and cover the grill, leaving the vents open, to smoke-roast them. Or you can begin by cooking them through over indirect heat and then coloring over direct flames. It's easy to cook sausages too fast, it's almost impossible to cook them too slowly—but generally they should take at least 10 to 15 minutes on the grill.

From a temperature standpoint, poaching is the gentlest way to cook sausage. If you place a sausage in water that's between 160 and 180 degrees F./71 and 82 degrees C., too hot to touch but not yet simmering—gentle, uniform, consistent—it will cook very uniformly.

Unfortunately, poaching is the one form of cooking that doesn't add any flavor to the sausage. So poaching should not be used for fresh sausages, but rather for those that already have exterior color and flavor–namely, smoked sausages, sausages that are already cooked, such as hot dogs. It should also be used for sausages that require very delicate heat for their stability, emulsified sausages such as weisswurst, which can be sautéed for color after being poached and chilled.

RULES TO LIVE BY WHEN MAKING SAUSAGE: A CONCISE PRIMER

1. Be sure your sausage includes at least 25 to 30 percent fat.
2. Use a precise amount of salt, measured on a scale.
3. Chill your meat in the freezer before grinding and mixing: the colder the better, just short of freezing (this is especially important when making sausage in a warm environment).
4. Always grind your meat into a bowl that is set in ice.
5. Mix the ground meat well to create a good, cohesive, not crumbly texture.
6. Adding ice-cold liquid when mixing the ground meat helps distribute the seasonings and results in a moister sausage.
7. Cook your sausage to an exact temperature, 150 degrees F./65 degrees C. in general, or, for sausages containing poultry, 160 degrees F./71 degrees C.
8. Savor every bite: A great sausage is a special creation.

SERVING

On a bun–that's among the most common ways to serve a sausage–and if you pile on some roasted peppers and onions, you've got three major food groups right there, a complete meal you can hold in your hand. Sausage is fantastic with pasta, of course, and loose sausage is perfect here, for those who like the idea of making sausage but don't have the time or tools to stuff it into casings. Moreover, making your own sausage for your pasta dishes allows you to choose your ingredients and seasonings to suit the dish. A chicken-basil-tomato sausage or a French garlic sausage might suit a pasta far better than a coil of store-bought "Italian."

Other starches work just as well. Merguez, a Moroccan spicy lamb sausage, goes great with couscous on its native turf. Bangers and mash are a classic combo in Britain; you can elevate that tradition with your own sausages and a smooth, buttery potato puree. Or, with rice as your backdrop, you can take sausage any direction you choose, from Asian to Spanish to American Southwestern cuisine, all depending on the seasonings you use.

Loose sausage can be cooked in many dishes—scrambled eggs, polenta, or a rice pilaf—or added to corn bread batter. Links or slices or chunks of sausage will transform a pot of beans or soups, flavoring the broth as they cook.

And of course many sausages, such as kielbasa and knackwurst, add incomparable flavor to braised sauerkraut, braised cabbage dishes, or any one-pot preparation in which the juices of the sausage can flavor the other ingredients.

Sausages also make the perfect hors d'oeuvre, whether a summer sausage sliced and served cold, or simply a fresh sausage served hot. People love sausage, it's easy to serve, and it's easy to eat standing at a party.

Sausages make the perfect snack—especially, believe it or not, if you're on a low-calorie diet. Because of its high fat content, one small sausage will keep hunger pangs away for hours. If you ever munch on rice cakes and carrot sticks, you know they don't quash hunger for long. The hard part, of course, is not eating too *many* sausages.

Even sausage fanatics don't usually think of sausage as a high-end item, something that would appear on the menu of a four-star restaurant, but often they do. At the French Laundry in California's Napa Valley, former sous-chef Eric Ziebold used to fashion a French sausage called a *cervelas*, speckled with pistachio and truffle, served sliced into disks with green lentils. Another of his dishes was a boudin blanc that was served with orange confit, fennel, and a pale-green pistachio vinaigrette. Bistro dishes elevated to four-star cuisine. Paul Bocuse, at his Michelin three-star restaurant outside Lyon, often serves a garlic sausage cooked en brioche as an *amuse-bouche*.

At Five Lakes Grill, Brian often makes sliced sausages a thematic component or garnish in a larger plate. He might serve a chicken sausage with a roasted chicken breast, a duck sausage with roast duck, venison sausage with loin of venison, or shrimp-lobster-leek sausage along with a grilled shrimp entrée.

Most fresh sausages, if they are made with fresh ingredients and well wrapped, will keep for 1 week in the refrigerator without a significant compromise in flavor. Sausages also freeze well, as do most high-fat items when well wrapped. Double-wrap them before freezing, or wrap them in plastic and store in freezer bags (keep a marker handy–it's wise to note the type of sausage and the date on the bag itself; don't think you'll remember). Sausages can be frozen for up to 3 months, depending on the freshness of the sausage, how well they are wrapped, and the temperature of your freezer, but they're at their peak if eaten fresh.

FRESH SAUSAGE MASTER RATIO

When you're making your own sausage, you may not always have precise amounts listed in various recipes, or you may simply want to invent your own recipes. These are the ratios we stick to when making sausage.

The meat-fat ratio should be about 70 percent meat and 30 percent fat.

Ratio:
100 percent meat and fat
1.75 percent (of the weight of the meat) salt
10 percent (of the weight of the meat) ice-cold liquid

Generally throughout this chapter we use the following:
5 pounds/2.25 kilograms meat and fat
3 tablespoons/40 grams kosher salt
Seasonings
1 cup/250 milliliters ice-cold liquid

FRESH SAUSAGE MASTER RECIPE:
FRESH GARLIC SAUSAGE

The master recipe for fresh sausage uses the ratio of 5 pounds/2.25 kilograms meat and fat to 1½ ounces/40 grams salt to 1 cup/250 milliliters liquid, which helps distribute seasonings, strengthens the bind, and enhances juiciness. We recommend that in most cases your meat-fat mixture include between 1 and 1½ pounds/450 and 675 grams fat. If you're using only pork shoulder butt, however, you may not need to add fat; it has 30 percent fat already, though this may depend on the exact cut; commercial pork butt is often sold very lean.

The type of liquid as well as additional seasonings all depend on the type of sausage you are making and your personal taste. The following sausage is simple but powerfully flavored with wine and garlic. Wine is an especially effective seasoning in sausage, adding an acidic balance to its fattiness.

> **5 pounds/2.25 kilograms boneless fatty pork shoulder butt,**
>   **diced**
> **3 tablespoons/40 grams kosher salt**
> **1 tablespoon/10 grams ground black pepper**
> **3 tablespoons/54 grams minced garlic**
> **1 cup/250 milliliters good red wine, chilled**
>
> **10 feet/3 meters hog casings, soaked in tepid water for at**
>   **least 30 minutes and rinsed**

1. Toss the meat, salt, pepper, and garlic together in a large bowl until evenly mixed. Cover and refrigerate until the mixture is thoroughly chilled, at least 2, and up to 24 hours. Alternatively, place in your freezer for 30 minutes to an hour, until the meat is very cold, even stiff, but not frozen.

2. Grind the mixture through the small die into a bowl set in ice (see Note below).

3. Using the paddle attachment of a standing mixer (or a strong wooden spoon if mixing by hand), mix on low speed (or stir) for 1 minute. Add the wine, increase the mixing speed to medium, and mix (or stir) for 1 more minute, or until the liquid is incorporated and the meat looks sticky.

4. Fry a bite-sized portion of the sausage and taste it (refrigerate the remaining sausage mixture while you do this and then set up your stuffing equipment). Adjust the seasoning if necessary and mix again to incorporate the additional seasoning.

5. Stuff the sausage into the hog casings and twist into 6-inch/15-centimeter links.

6. Cook the sausage to an internal temperature of 150 degrees F./65 degrees C.

Yield: About 5 pounds/2.25 kilograms sausage;
about twenty 6-inch/15-centimeter links

[NOTE: See pages 105–115 for a detailed description of the basic grinding, mixing, stuffing, and cooking techniques.]

BRIAN'S HOLIDAY KIELBASA, WIEJSKA (KIELBASA WITH MARJORAM)

"This was my introduction to charcuterie, so it's a very personal recipe," Brian says. "This is what the family ate every Christmas and Easter. Mom used to take all the food to the priest to have it blessed–her deference to the magic element in charcuterie. The real key is to buy good pork. Find a butcher you like, whether at a specialty market or at a grocery store, and develop a relationship with him or her. The second key is to toss the ingredients together well and refrigerate them overnight to allow the seasoning to develop and disperse.

"Mom's instinct is to trim away some of the fat because she's watching her fat intake, but I leave it on so the kielbasa will be more juicy. Also, Mom puts the whole coil on a baking sheet, adds water, and in effect steams the sausage in the oven, then slices it to serve. I think this sausage benefits from dry-heat cooking–that is, sautéing or roasting–rather than moist heat (steaming or poaching), though a combination of the two, a quick braise, also works well.

"Mom serves this with her pierogies–big dumplings loaded with hot melted cheese. A platter of kielbasa and pierogies, that is the meal. Because the priest blesses it, we have to eat every morsel, which isn't a problem, but it doesn't leave a lot of room for vegetables!"

5 pounds/2.25 kilograms boneless shoulder butt, diced

3 tablespoons/40 grams kosher salt

¼ cup/72 grams minced garlic

3 tablespoons/18 grams coarsely chopped fresh marjoram (or

1½ tablespoons/3 grams ground dried marjoram)

1 tablespoon/10 grams ground black pepper

½ cup/125 milliliters ice water

10 feet/3 meters hog casings, soaked in tepid water for at
least 30 minutes and rinsed

1. Combine all the ingredients except the water and toss to distribute the seasonings. Refrigerate overnight.

2. Grind the mixture through the small die into a bowl set in ice (see Note below).

3. Add the water to the meat, mixing with the paddle attachment (or a sturdy spoon) until the water is incorporated and the mixture has developed a uniform, sticky appearance, about 2 minutes on medium speed.

4. Sauté a small portion of the sausage, taste, and adjust the seasoning if necessary.

5. Stuff the sausage into the hog casings and twist into 6-inch/15-centimeter links. Refrigerate or freeze until ready to cook.

6. Gently sauté or roast the sausage to an internal temperature of 150 degrees F./65 degrees C.

Yield: About 5 pounds/2.25 kilograms sausage;
about twenty 6-inch/15-centimeter links

[NOTE: See pages 105–115 for a detailed description of the basic grinding, mixing, stuffing, and cooking techniques.]

BREAKFAST SAUSAGE WITH FRESH GINGER AND SAGE

Homemade breakfast sausage (aka Da Bomb) is one of the easiest sausages to make, tastes far better than store-bought, and doesn't require stuffing if you don't have the materials or inclination–it can be shaped into patties and cooked that way. It can also be rolled into a log, wrapped in plastic, and frozen, and disks can be cut as needed.

Fresh ginger makes this sausage vibrant and fresh sage enhances the clarity of the flavors. This sausage is best sautéed so that its aroma fills a morning kitchen, but it can be roasted. It's also excellent grilled or smoked (see page 74 for smoking techniques).

5 pounds/2.25 kilograms boneless pork shoulder butt, diced

3 tablespoons/40 grams kosher salt

5 tablespoons/50 grams peeled and finely grated fresh ginger
    (or 1 tablespoon/8 grams ground dried ginger)

5 tablespoons/30 grams tightly packed finely chopped
    fresh sage

1 tablespoon/18 grams minced garlic

2 teaspoons/6 grams ground black or white pepper

1 cup/250 milliliters ice water

20 feet/6 meters sheep casings or 10 feet/3 meters hog
    casings, soaked in tepid water for at least 30 minutes
    and rinsed (optional)

1. Combine all the ingredients except the water and toss to distribute the seasonings. Chill until ready to grind.

2. Grind the mixture through the small die into a bowl set in ice (see Note below).

3. Add the water to the meat mixture and mix with the paddle attachment (or a sturdy spoon) until the liquid is incorporated and the mixture has developed a uniform, sticky appearance, about 1 minute on medium speed.

4. Sauté a small portion of the sausage, taste, and adjust the seasoning if necessary.

5. Stuff the sausage into the casings and twist into 4-inch/10-centimeter links, or shape into patties; refrigerate or freeze until ready to cook; or roll into a log, wrap in plastic and freeze, slice into patties.

6. Gently sauté or roast the sausage to an internal temperature of 150 degrees F./65 degrees C.

Yield: About 5 pounds/2.25 kilograms sausage;
about sixty 4-inch/10-centimeter links

[NOTE: See pages 105–115 for a detailed description of the basic grinding, mixing, stuffing, and cooking techniques.]

CLASSIC FRESH BRATWURST

The ultimate fresh bratwurst, this is one of the richest sausages here, given its generous use of cream and eggs. And because of the additional dairy fat, which needs to be emulsified (uniformly distributed) into the mixture, keeping your ingredients very cold is especially important. (If the sausage "breaks," the flavor remains but the texture will be unpalatable.) This is a big juicy sausage made from pork and veal with the traditional sweet-spice bratwurst flavors of nutmeg and ginger.

3 pounds/1350 grams boneless pork shoulder butt, diced

1 pound/450 grams boneless lean veal shoulder, diced

1 pound/450 grams pork back fat

3 tablespoons/40 grams kosher salt

2 teaspoons/6 grams ground white pepper

1½ teaspoons/5 grams ground ginger

1½ teaspoons/5 grams freshly grated nutmeg

2 large cold eggs, lightly beaten

1 cup/250 milliliters ice-cold heavy cream

10 feet/3 meters hog casings, soaked in tepid water for at least 30 minutes and rinsed

1. Combine all the ingredients except the eggs and cream and toss well to distribute the seasonings. Chill until ready to grind.

2. Grind the mixture through the small die into a bowl set in ice (see Note below).

3. Using the paddle attachment of a standing mixer (or a strong wooden spoon if mixing by hand), mix on low speed (or stir) for 1 minute. Add the eggs and cream, start the mixer on low, and then increase the speed to medium and mix until the cream and eggs are uniformly incorporated and the sausage appears sticky, about a minute longer. Sauté a small portion of the sausage and taste; adjust the seasoning if necessary. (Refrigerate the sausage mixture while you do this.)

4. Stuff the sausage into the hog casings. Twist into 6-inch/15-centimeter links. Refrigerate or freeze until ready to cook.

5. Gently sauté or roast the sausage to an internal temperature of 150 degrees F./65 degrees C.

Yield: About 5 pounds/2.25 kilograms sausage;
about twenty 6-inch/15-centimeter links

[NOTE: See pages 105–115 for a detailed description of the basic grinding, mixing, stuffing, and cooking techniques.]

ITALIAN SAUSAGE, SPICY AND SWEET

Italian sausage is an everyday-style sausage, inexpensive ingredients simply combined and heavily seasoned. This is an excellent version of what is so-often pedestrian sausage, the kind the supermarket makes with its pork scraps. While the sausage does have roots in a sausage made in Italy, in the United States its name is more or less a generic term for a heavily seasoned pork sausage, usually including fennel or anise, seasonings associated with the early Italian immigrants or popularized by immigrants from the southern part of the country. Ours also includes paprika, oregano, and basil and is finished with red wine vinegar. If you like it very hot, increase the red pepper flakes in the first version by 50 percent; if you don't like chile peppers, make the sweet version, "sweet" indicating not the presence of sweetness but rather the absence of heat. A powerfully seasoned sausage such as this one is often used as a component in a big minestrone. It does go well with pasta. Sautéed onions and peppers, with their sweetness, mix beautifully with these flavors.

FOR SPICY SAUSAGE

4½ pounds/2 kilograms boneless pork shoulder butt, diced

8 ounces/225 grams pork back fat, diced

3 tablespoons/40 grams kosher salt

2 tablespoons/32 grams granulated sugar

2 tablespoons/16 grams fennel seeds, toasted (see page 51)

1 tablespoon/8 grams coriander seeds, toasted (see page 51)

3 tablespoons/24 grams Hungarian paprika

½ teaspoon/1 gram cayenne pepper

4 tablespoons/24 grams fresh oregano leaves

4 tablespoons/24 grams fresh basil leaves

2 tablespoons/12 grams hot red pepper flakes

2 teaspoons/6 grams coarsely ground black pepper

¾ cup/185 milliliters ice water

¼ cup/60 milliliters red wine vinegar, chilled

FOR SWEET SAUSAGE

4 pounds/1800 grams boneless pork shoulder butt, diced into
    1-inch pieces
1 pound/450 grams pork back fat, diced into 1-inch pieces
3 tablespoons/40 grams kosher salt
2 tablespoons/32 grams granulated sugar
2 teaspoons/12 grams minced garlic
2 tablespoons/16 grams fennel seeds, toasted (see page 51)
2 teaspoons/6 grams coarsely ground black pepper
2 tablespoons/16 grams sweet Spanish paprika
¾ cup/185 milliliters ice water
¼ cup/60 milliliters red wine vinegar, chilled

10 feet/3 meters hog casings, soaked in tepid water for at
    least 30 minutes and rinsed

1. Combine all the ingredients except the water and vinegar and toss to distribute the seasonings. Chill until ready to grind.

2. Grind the mixture through the small die into a bowl set in ice (see Note below).

3. Add the water and vinegar to the meat mixture and mix with the paddle attachment (or a sturdy spoon) until the liquids are incorporated and the mixture has developed a uniform, sticky appearance, about 1 minute on medium speed.

4. Sauté a small portion of the sausage, taste, and adjust the seasoning if necessary.

5. Stuff the sausage into the hog casings, and twist into 6-inch/15-centimeter links. Refrigerate or freeze until ready to cook.

6. Gently sauté or roast the sausage to an internal temperature of 150 degrees F./65 degrees C.

Yield: About 5 pounds/2.25 kilograms sausage;
about twenty 6-inch/15-centimeter links

[NOTE: See pages 105–115 for a detailed description of the basic grinding, mixing, stuffing, and cooking techniques.]

CHICKEN SAUSAGE WITH BASIL AND TOMATOES

This recipe, one of our favorites in the book, reaches into the realm of the artisanal-style sausage because of the balance of ingredients and the complexity of flavors (this uses vinegar to accentuate them, as well as an additional fat for seasoning in the form of flavorful olive oil). But all it really does is combine flavors we all know go great together. If you were to combine the seasonings listed here, puree them, and then sauté, that would work beautifully as a sauce for grilled chicken. All sausages can be invented this way.

This sausage is particularly good gently grilled, but it can be roasted or sautéed. Serve it traditionally, as a sandwich or sliced with mustard, but look again at the ingredients—it would work beautifully, in casings or loose, with a simple pasta, along with garlic, oil, and Parmigiano-Reggiano.

3 ½ pounds/1.5 kilograms boneless, skinless chicken thighs, cubed

1½ pounds/675 grams pork back fat, cubed, diced into 1-inch pieces (see Note 1 below)

3 tablespoons/40 grams kosher salt

1 teaspoon/3 grams freshly ground black pepper

1½ teaspoons/9 grams minced garlic

4 tablespoons/24 grams tightly packed chopped fresh basil

½ cup/100 grams fresh diced Roma (plum) tomatoes

¼ cup/60 grams diced sun-dried tomatoes (see Note 2 below)

¼ cup/60 milliliters red wine vinegar, chilled

¼ cup/60 milliliters extra virgin olive oil

¼ cup/60 milliliters dry red wine, chilled

10 feet/3 meters hog casings, soaked in tepid water for at least 30 minutes and rinsed

1. Combine the meat, fat, salt, pepper, garlic, basil, and tomatoes and toss together until evenly mixed. Chill until ready to grind.

2. Grind the mixture through the small die into a bowl set in ice (see Note 3 below).

3. Using the paddle attachment of a standing mixer (or a sturdy spoon), mix on low speed (or stir) for 1 minute. Add the vinegar, oil, and wine, increase the speed to medium, and mix for 1 more minute, or until the liquid is incorporated and the sausage has a uniform, sticky appearance.

4. Fry a bite-sized portion of the sausage, taste and, adjust the seasoning if necessary.

5. Stuff the sausage into the hog casings and twist into 6-inch/15-centimeter links. Refrigerate or freeze until ready to cook.

6. Cook the sausage to an internal temperature of 160 degrees F./71 degrees C.

Yield: About 5 pounds/2.25 kilograms sausage;
about twenty 6-inch/15-centimeter links

[NOTES: 1. To make this without pork, omit the pork fat but retain as much chicken fat on the thighs as possible. 2. Any sun-dried tomatoes will work as long as they are pliable. Those not packed in oil or some kind of liquid, though, may be hard. If this is the case, pour boiling water over them and allow to soak until soft. 3. See pages 105–115 for a detailed description of the basic grinding, mixing, stuffing, and cooking techniques.]

DUCK, SAGE, AND ROASTED GARLIC SAUSAGE

This is a main-course sausage. If you're going to go to the effort and expense of using duck to create a truly unusual and exquisite sausage, it should be the centerpiece of the meal. Actually, this recipe is really no more difficult than any other fresh sausage (and easier if you ask your butcher to bone the ducks for you, saving all the skin and bones and fat for other uses), but when you serve it to guests, it makes them feel special because it's rare sausage.

3½ pounds/1.5 kilograms raw duck meat (preferably from
the legs and thighs), trimmed of all fat, sinew, and skin
and diced, 8 to 10 Pekin legs and thighs
1½ pounds/675 grams pork back fat, diced
3 tablespoons/40 grams kosher salt
½ cup/64 grams finely chopped fresh sage

3 tablespoons/60 grams Steam-Roasted Garlic Paste (recipe
follows)
1½ tablespoons/15 grams coarsely ground black pepper
½ cup/125 milliliters ice water
½ cup/125 milliliters dry red wine, chilled

10 feet/3 meters hog casings, soaked in tepid water for at
least 30 minutes and rinsed

1. Combine all the ingredients except the water and wine and toss to distribute the seasonings. Chill until ready to grind.

2. Grind the mixture through the small die into a bowl set in ice (see Note below).

3. Add the water and wine to the meat mixture and mix with the paddle attachment (or
sturdy spoon) until the liquids are incorporated and the mixture has developed a uniform,
sticky appearance, about 1 minute on medium speed.

4. Sauté a small portion of the sausage, taste, and adjust the seasoning if necessary.

5. Stuff the sausage into the hog casings and twist into 6-inch/15-centimeter links.
Refrigerate or freeze until ready to cook.

6. Gently sauté or roast the sausage to an internal temperature of 150 degrees F./65
degrees C.

Yield: About 5 pounds/2.25 kilograms sausage;
about twenty 6-inch/15-centimeter links

[NOTE: See pages 105–115 for a detailed description of the basic grinding, mixing, stuffing, and cooking techniques.]

Steam-Roasted Garlic Paste

Roasted garlic is a fantastic all-purpose seasoning for sausages, sauces, soups,
and potatoes and other starches—it's hard to think of much that wouldn't be improved with
some roasted garlic. Steam-roasting the garlic avoids the deeper, sometimes harsh, flavors
that can develop when garlic is simply roasted in foil. Garlic prepared this way is both paler
and sweeter than garlic roasted at higher temperatures.

The roasted garlic will mash easily, but it's a good idea to press it through a sieve to remove any remaining pieces of skin.

6 whole heads garlic

1. Preheat the oven to 325 degrees F./160 degrees C.
2. Place the garlic root side down in a baking pan just large enough to hold it. Add water to come to a depth of ¼ inch/0.5 centimeter and cover with foil. Bake until the garlic is completely soft, about 1 hour. Let cool.
3. Cut heads of garlic horizontally in half and squeeze the soft pulp into a sieve set over a bowl. Push the garlic through the sieve with a rubber spatula. Store in the refrigerator, covered, for up to a week.

Yield: About ¾ cup/185 milliliters

MEXICAN CHORIZO

I find this to be among the most exotic tasting sausages because of the sweet ancho powder and the hot, smoky chipotle powder, offset by the fresh oregano and garlic. Although this can be stuffed into casings, Mexican chorizo, unlike Spanish chorizo, is generally a free-form sausage used as a component in other dishes, rather than eaten by itself like a breakfast patty. It's excellent with scrambled eggs (cook the sausage first, then add the eggs, or cook the eggs separately and add the cooked sausage and juices to them). It's delicious in a corn tortilla; stuffed into a poblano pepper and roasted, or stuffed into a roasted poblano, battered, and deep-fried; sautéed with black beans and red onion; or added to corn bread or polenta.

5 pounds/2.25 kilograms boneless pork shoulder butt, diced
3 tablespoons/40 grams kosher salt
2 tablespoons/16 grams ancho chile powder (see Note 1 below)
1 tablespoon/8 grams hot paprika
1 tablespoon/8 grams chipotle powder (see Note 1 below) or cayenne powder

1 tablespoon/18 grams minced garlic

1 teaspoon/3 grams freshly ground black pepper

1 tablespoon/6 grams chopped fresh oregano (or 1
teaspoon/0.5 gram dried oregano)

½ teaspoon/1.5 grams ground cumin

3 tablespoons/45 milliliters tequila, chilled

3 tablespoons/45 milliliters red wine vinegar, chilled

10 feet/3 meters hog casings, soaked in tepid water for at
least 30 minutes and rinsed

1. Combine all the ingredients except the tequila and vinegar and toss to distribute the seasonings. Chill until ready to grind.

2. Grind the mixture through the small die into a bowl set in ice (see Note 2 below).

3. Add the tequila and vinegar to the meat mixture and mix with the paddle attachment (or a sturdy spoon) until the liquids are incorporated and the mixture has developed a uniform, sticky appearance, about 1 minute on medium speed.

4. Sauté a small portion of the sausage, taste, and adjust the seasoning if necessary.

5. If desired: stuff into the hog casings and twist into 6-inch/15-centimeter links. Refrigerate or freeze until ready to cook.

6. If using links, gently sauté or roast to an internal temperature of 150 degrees F./65 degrees C. If using loose, sauté until cooked through.

Yield: About 5 pounds/2.25 kilograms sausage;
about twenty 6-inch/15-centimeter links

[NOTES: 1. Generic chili powder can be used, but ancho chiles (dried poblanos) have a distinctive sweet, spicy flavor. These chile powders can be found at supermarkets, at specialty stores, and through mail-order. To make your own powder, roast the dried chiles in a 300-degree-F. /150-degree-C. oven until they are very hard and dry, about 15 minutes. Allow them to cool to room temperature, then break them open, discard the seeds and stems, and pulverize in a coffee or spice grinder. 2. See pages 105–115 for a detailed description of the basic grinding, mixing, stuffing, and cooking techniques.]

MERGUEZ

A spicy lamb sausage with North African roots, Merguez is popular there and in France. It's got a distinctive, lamby flavor and a bright red color from the paprika, as well as extraordinary juiciness from the roasted red pepper. Serve with couscous and sautéed bell peppers.

4 pounds/2 kilograms boneless lamb shoulder, diced

1 pound/450 grams pork back fat, diced

3 tablespoons/40 grams kosher salt

2 teaspoons/5 grams sugar

1 teaspoon/2 grams hot red pepper flakes

2 tablespoons/18 grams minced garlic

1½ cups/175 grams diced roasted red peppers

1½ teaspoons/5 grams freshly ground black pepper

2 tablespoons/16 grams Spanish paprika

2 tablespoons/16 grams minced fresh oregano

¼ cup/60 milliliters dry red wine, chilled

¼ cup/60 milliliters ice water

20 feet/6 meters sheep casings, soaked in tepid water for at least 30 minutes and rinsed

1. Combine all the ingredients except the wine and water and toss to distribute the seasonings. Chill until ready to grind.

2. Grind the mixture through the small die into a bowl set in ice (see Note below).

3. Add the wine and water to the meat mixture and mix with the paddle attachment (or a sturdy spoon) until the liquids are incorporated and the mixture has developed a uniform, sticky appearance, about 1 minute on medium speed.

4. Cook a small portion of the sausage, taste, and adjust the seasoning if necessary.

5. Stuff the sausage into the sheep casings and twist into 10-inch/25-centimeter links. Refrigerate or freeze until ready to cook.

6. Gently sauté or roast the sausage to an internal temperature of 150 degrees F./65 degrees C.

Yield: About 5 pounds/2.25 kilograms sausage; about twenty-four 10-inch/25-centimeter links

[NOTE: See pages 105–115 for a detailed description of the basic grinding, mixing, stuffing and cooking techniques.]

TO ROAST BELL PEPPERS

Raw bell peppers are rather pedestrian–a lot of fiber, a lot of water, not a lot of taste. Roast them, though, and they are absolutely transformed. Intensely flavored and wonderfully sweet, they become another superb all-purpose flavoring for pastas and salads, a great garnish for chicken or fish, or even the star of a dish–in, say, a roasted red pepper soup.

Raw bell peppers

1. Roast the peppers over an open flame on a gas stove, on a grill over very hot coals, or under the broiler, turning frequently, until the skins are completely charred. Place them in a bowl, cover with plastic wrap, and let sit for at least 30 minutes.

2. Remove the charred black outer skin from the peppers. Cut them in half lengthwise and remove the seeds and cores. Pat dry with paper towels. Store in a covered container in the refrigerator for up to a week.

SPICY ROASTED POBLANO SAUSAGE

This is another sausage that uses roasted peppers to impart an intriguing flavor as well as great juiciness. Brian has served it with sautéed shrimp, green chile rice, and a black bean sauce; alongside a grilled chicken breast with guacamole, Jack cheese, and a tomato coulis; and in a tortilla with slaw and spicy tomatoes. This sausage, no more difficult to make than Italian sausage, will surprise guests with its unexpected poblano flavor.

5 pounds/2.25 kilograms boneless pork shoulder butt, diced

3 tablespoons/40 grams kosher salt

3 tablespoons/24 grams ancho chile powder (see Note 1 below)

2 teaspoons/6 grams ground cumin

2 tablespoons/12 grams fresh oregano

1 tablespoon/18 grams minced garlic

1 tablespoon/8 grams Spanish paprika

1 cup/120 grams small-diced roasted, peeled, and seeded
    poblano peppers

8 tablespoons/48 grams chopped fresh cilantro

1 cup/250 milliliters ice water

10 feet/3 meters hog casings, soaked in tepid water for at
    least 30 minutes and rinsed

1. Combine all the ingredients except the poblano peppers, cilantro, and water, toss to distribute the seasonings. Chill until ready to grind.

2. Grind the mixture through the small die into a bowl set in ice (see Note 2 below).

3. Add the peppers and cilantro to the meat mixture. Mix with the paddle attachment (or a sturdy spoon) while slowly adding the water. Continue mixing until all the liquid is incorporated and the sausage has developed a uniform, sticky appearance, about 1 minute on medium speed.

4. Sauté a small portion of the sausage, taste, and adjust the seasoning if necessary.

5. Stuff the sausage into the hog casings and twist into 6-inch/15-centimeter links. Refrigerate or freeze until ready to cook.

6. Sauté or roast the sausage to an internal temperature of 150 degrees F./65 degrees C.

Yield: About 5 pounds/2.25 kilograms sausage;
about twenty 6-inch/15-centimeter links

[NOTES: 1. Generic chili powder can be used, but ancho chiles (dried poblanos) have a distinctive sweet spicy flavor. Ancho chile powder can be found at specialty stores, at many supermarkets, and via mail-order. To make your own powder, roast the anchos in a 300-degree-F./150-degree-C. oven until they are very hard and dry, about 15 minutes. Allow them to cool to room temperature, then break them open, discard the seeds and stems, and pulverize in a coffee or spice grinder. 2. See pages 105–115 for a detailed description of the basic grinding, mixing, stuffing, and cooking techniques.]

TURKEY SAUSAGE WITH DRIED TART CHERRIES

The pronounced turkey flavor of this sausage, which uses the abundant dark meat of a turkey, is offset by the sweet-tart cherries. If you can get high-quality farm-raised turkeys, you might want to brine and smoke the breast and then use the dark meat and fat for the sausage. If your turkey has abundant fat, use as much of that as possible in place of some of the pork fat. This sausage can be served on its own, but it also works nicely as an impressive, and appropriate, accompaniment to grilled turkey breast.

1 cup/200 grams dried tart cherries, picked through to remove
    any stray pits
1/2 cup/125 milliliters dry white wine
3 1/2 pounds/1.5 kilograms skinless, boneless turkey leg and
    thigh meat, cartilage and sinew removed, diced
1 1/2 pounds/675 grams pork back fat (or a combination of pork
    and turkey fat), diced
3 tablespoons/40 grams kosher salt
2 tablespoons/20 grams sugar
2 teaspoons/6 grams ground white pepper
2 tablespoons/12 grams minced fresh sage
1/2 teaspoon/1 gram grated fresh ginger
1/2 teaspoon/2 grams ground cloves
1/4 teaspoon/1 gram ground cinnamon

10 feet/3 meters hog casings, soaked in tepid water for at
    least 30 minutes and rinsed

1. Soak the dried cherries in the wine for 30 minutes. Drain the cherries, and reserve the wine. Refrigerate the wine to chill it.

2. Combine all the ingredients except the wine and toss to distribute the seasonings. Chill until ready to grind.

3. Grind the mixture through the small die into a bowl set in ice (see Note below).

4. Add the wine to the meat mixture and mix with the paddle attachment (or sturdy spoon) until the liquid is incorporated and the mixture has developed a uniform, sticky appearance, about 1 minute on medium speed.

5. Cook a small portion of the sausage, taste, and adjust the seasoning if necessary.

6. Stuff the sausage into the hog casings and twist into 6-inch/15-centimeter links. Refrigerate or freeze until ready to cook.

7. Gently sauté or roast the sausage to an internal temperature of 160 degrees F./71 to 70 degrees C.

Yield: About 5 pounds/2.25 kilograms sausage;
about twenty 6-inch/15-centimeter links

[NOTE: See pages 105–115 for a detailed description of the basic grinding, mixing, stuffing, and cooking techniques.]

Emulsified Sausages

An emulsified sausage is one in which the seasoned meat and fat are pureed in a food processor or mixed to death in a standing mixer, until the fat is evenly suspended throughout the meat and water. The meat-fat-water emulsion looks like a smooth, creamy paste. When stuffed into casings and cooked, it has a superfine texture and uniform color. Think of the inside of a hot dog, one of the most familiar emulsified sausages. Or compare a slice of bologna, which is emulsified, with a slice of salami, which is not.

Making an emulsified sausage involves a few more steps than making fresh sausage.

Three different types of sausage (from left to right): Hunter sausage, which is smoked; an emulsified sausage with chunks of ham inside; and Spanish chorizo, a dry-cured sausage.

The process is not difficult, but the temperature must be watched vigilantly to ensure a stable emulsion and a delicious sausage. Just as hollandaise, a classic emulsion sauce, can break, so too can an emulsion sausage mixture. A broken sausage emulsion is sadder by far than a broken hollandaise, and there's little to be done about it. And you won't necessarily be able to see it happening. Sometimes it will break precipitously, turning into the consistency of oatmeal, but another time, you might only notice that it has become very shiny on the surface. That may in fact indicate a broken emulsion, but you won't know it until you take your first bite. Though flavorful, it will feel a little like eating clumps of soggy ground-up newspaper.

Done properly, though, making an emulsified sausage is an exciting process, a test of real craftsmanship in your manipulation of protein and fat, in the transformation of chunks of meat and fat and seasonings into a highly refined and flavorful package. Crafty though it is in its alchemy, it's not difficult to do as long as you follow the directions here to a tee. Then it's a breeze. If you serve one of these sausages to friends at home, they're apt to look at you and say, "You made that? How did you make it?" Most people aren't used to seeing homemade emulsified sausages, and so the creations that follow are especially impressive.

Yet, because their texture tends to be so fine and light, they lend themselves just as well to being served on a bun as featured in a formal first course, simply sliced on their own with some sauce or mustard. Homemade emulsified sausages tend to be much lighter on the palate than commercial emulsified sausages, such as hot dogs, which are denser and usually almost rubbery in texture (this is due mainly to the speed with which industrial food processors emulsify and the pressure with which the casings are stuffed, as well as to additives designed to prolong shelf life). Homemade emulsified sausages are very light on the palate, very delicate, almost creamy, like a mousseline–and they should be treated accordingly.

Emulsified sausages are usually poached very gently till fully cooked, then quickly chilled (this cooking solidifies the emulsion and to some extent helps to preserve it). They're often smoked for flavor and color. Smoking turns the casing a tantalizing golden brown. Once they're cooked or smoked, they can then be reheated however you like: roasted, grilled, gently sautéed, poached again. In any case, they should be eaten within several days. They also keep well frozen for up to six weeks, after which the quality of the flavor begins a slow decline.

We're offering two different methods for making emulsified sausages, using a standing mixer with the paddle attachment or a food processor. We prefer the standing mixer because it results in sausage with a firmer texture. The blades of most home food processors are not as sharp as they need to be for this technique. By the time the processor has created the emulsion, the blade will have thoroughly whipped the meat, overaerating it as it were. The result is a stable sausage but the texture is very delicate. However, both methods work.

The particular emulsion we use here is called a 5-4-3 forcemeat (a ground mixture used for stuffing), from the French term (*farcir*) for its ratio of meat to fat to water. If you remember that, and the 1.75 percent salt, you almost don't need a recipe. The same general sausage-making rules as far as keeping all your equipment cold, and all your moist ingredients chilled apply only more so. It's prudent to put your grinder, the mixing bowl, and the paddle attachment or food processor blade in the freezer for an hour or so before you begin. The mechanical act of pureeing or mixing, the friction, will heat the ingredients.

Basic Emulsified Sausage: Standing Mixer Method

Fritz Sonnenschmidt, an exuberant chef, raconteur, teacher of legions before he retired from the Culinary Institute of America, author of a book on the craft of garde-manger, and a kind of charcuterie godfather in this country, offered this method for making emulsified sausages.

TOOLS

You'll need basically the same tools for emulsified sausage as for fresh sausage (see the list on page 105). But Brian and I don't recommend using the stuffer attachment for the mixer because of the risk of the mixture's breaking and the mess of feeding a paste through the tube.

CHILLING

Both the grinder, fitted with blade and die, and the bowl and paddle of the mixer should be put in the freezer for an hour before you begin. The meat and fat should be partially frozen after they are diced. Very cold temperatures are critical in forming the emulsion.

THE FIRST GRIND

Remove the grinder from the freezer. Grind the meat and fat through the large die onto a chilled baking sheet. Store in the freezer until ready to use. The general rule for everything here is the colder, the better. The meat can go as far as crunchy on the exterior–but it shouldn't be frozen solid.

EMULSIFIED SAUSAGE

To make an emulsified sausage, one in which fat and meat are uniformly dispersed in a finely textured sausage, the ice-cold fat and meat are ground a second time with ice, then mixed with the paddle attachment of a standing mixer for several minutes until a heavy paste forms.

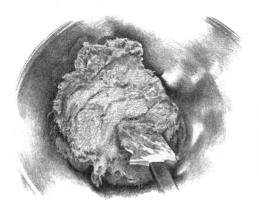

The emulsified meat should have a smooth, uniform texture. Any interior garnish, such as finely diced ham, should be folded into the sausage mixture at this stage.

SEASONING

Season the ground meat and fat with the salt (as well as sugar and/or pink salt in recipes that use them) after they have chilled.

THE SECOND GRIND

Remove the ground meat and fat from the freezer. Combine with the crushed ice and grind through the small die into the mixer bowl, set in ice.

EMULSIFYING

Remove the paddle attachment from the freezer and fit it onto the mixer. Add the ground meat mixture and the remaining seasonings to the bowl and mix on high speed for 3 to 4 minutes. Scrape down the sides of the bowl. The mixture should begin to look like a very stiff batter. Mix on high for 2 more minutes. Refrigerate the mixture in the mixing bowl until you're ready to stuff the casings.

THE QUENELLE TEST

The quenelle test provides an idea of what the finished sausage will taste like, while there's still time to tweak it if desired. Wrap a small bit of the mixture in plastic wrap (be sure to keep the rest of the mixture cold during your tests), roll it up tightly, and tie both ends; it should be about the size of your thumb, just big enough to check the seasoning and the texture and firmness. Drop the quenelle into water that's between 160 to 180 degrees F./71 to 82 degrees C, just below a simmer. When the interior of the quenelle reaches 150 degrees F./65 degrees C. (stick the thermometer through the plastic wrap) unwrap and taste it. Is it seasoned properly? If not, add more salt or appropriate seasonings to the mixture and mix briefly to distribute the seasonings. If your quenelle has an uncomfortably mealy texture, however, or if your plastic wrap is filled with melted fat, you have a broken forcemeat on your hands: excellent nutrition for pets.

GARNISHING

Many emulsified sausages contain chunks of meat, fat and/or nuts, what professional chefs refer to as "interior garnish." For example, consider using chunks of smoked ham or smoked tongue in a pork sausage. The solid chunks of meat in the finely textured sausage add great visual appeal, flavor, and another texture. Don't hesitate to use the same sort of garnish you might find in a pâté for an appropriate sausage, such as chunks of cooked mushrooms or fresh herbs, for visual impact and more flavor.

STUFFING

The technique is the same as for fresh sausage (see page 109), but we do not recommend using the KitchenAid stuffer attachment because of the risk of the mixture's breaking and the mess of feeding a paste through the tube.

Instead of stuffing the sausage into casings, you can pipe it out of a pastry bag directly into hot poaching liquid (preferably stock, but water will work).

Bring a large pot of water to a temperature between 160 and 180 degrees F./71 to 82 degrees C. Add the sausages. They will want to float, so place some sort of weight on top to keep them submerged and ensure even cooking: if your pot has a steamer insert, this works nicely; a plate or smaller lid will also work. Monitor the temperature of the sausages from time to time with an instant-read thermometer. Meanwhile, prepare an ice bath in a large bowl or pot.

When the interior temperature of the sausages reaches 150 degrees F./65 degrees C., about 15 to 20 minutes, remove them from the pot and submerge them in the ice bath to halt the cooking and chill them thoroughly (about 10 minutes, or to a temperature of 60 degrees F./15 degrees C.).

Some sausages are smoked first and then poached to finish the cooking. This section includes emulsified sausages that are not smoked. Smoked emulsified sausages are included in the final section of this chapter.

Once the sausages are poached and chilled, they can be reheated in a pan, in the oven, or on a grill, or poached again. But they are fully cooked at this stage, and you could safely eat them cold if you wished. The reason you reheat them is to give their skin a little more flavor through caramelization in the sauté pan or oven, or on the grill, and for overall flavor and succulence.

Basic Emulsified Sausage: Food Processor Method

TOOLS

You'll need basically the same tools for emulsified sausage as for fresh sausage, with the addition of a food processor with a standard 11-cup bowl or larger; see the list on page 105. But Brian and I don't recommend using the stuffer attachment for the mixer because of the risk of the mixture's breaking and the mess of feeding a paste through the tube. If you don't have a stuffer, you can pipe the sausage directly into heated water or stock using a canvas pastry bag.

CHILLING

Place the grinder, mixing bowl and food processor bowl and blade in the freezer for 1 hour before using.

SEASONING

Season only the meat with the salt (as well as sugar and/or pink salt in recipes using them) as you would for fresh sausage. Do not season the fat, which should be kept separate from the meat.

GRINDING

Remove the grinder from the freezer. Grind the meat and fat into separate bowls set in ice. Store in the freezer until ready to mix. The meat can go as far as crunchy on the exterior but should not be frozen solid.

EMULSIFYING

In order to ensure a stable emulsion, you must keep the sausage mixture at the temperatures recommended here. It's the best way to prevent winding up with broken sausage; there's not much you can do with a broken sausage emulsion. (These formulas, based on those used in the commercial meat industry, are taught at the Culinary Institute of America.)

Add the ground meat to the food processor. Next add the crushed ice, then the remaining seasonings. Process until the mixture is smooth and the ice has been thoroughly incorporated, about 2 minutes. Check the temperature with an instant-read thermometer; the meat should be below freezing. (Typically it will be at about 25 to 28 degrees F./−3 to −4 degrees C. at this point. If it's not below freezing, immediately spread it on a baking sheet and put it in the freezer until it reaches 32 degrees F./0 degrees C.)

Continue to process the meat until the temperature, which will be rising due to the friction of the whipping blade and the ambient temperature, reaches 40 degrees F./4 degrees C. This will take about 5 minutes (and will seem longer); take the meat's temperature with the thermometer every few minutes. When it reaches 40 degrees F./4 degrees C., add the ground fat and continue to puree. Depending on how cold your fat is, the temperature may drop again; this is good and will result in a smoother texture.

Continue to puree the mixture for a few more minutes, but don't let it get hotter than 58 degrees F./14 degrees C. At around 60 degrees F./15 degrees C., the fat will separate from the meat and liquids, so 58 degrees F./14 degrees C. is as close as you can safely take it.

Refrigerate the sausage in the food processor bowl till you're ready to stuff it into the casings.

Do a quenelle test as described on page 136 and then proceed as in the standard mixer method.

EMULSIFIED SAUSAGE MASTER RECIPE: WEISSWURST

This is a classic 5-4-3-style forcemeat in which the fat equals 50 percent of the total meat and water, resulting in a delicate, finely textured sausage. Here we use a standing mixer to paddle the sausage but you can also use a food processor as described on pages 137–138. This weisswurst is very juicy, and made vibrant by the lemon zest and fresh herbs.

1 pound/450 grams boneless lean veal shoulder, diced

12 ounces/350 grams pork back fat, diced

1½ tablespoons/20 grams kosher salt

8 ounces/225 grams crushed ice

1 teaspoon/3 grams ground white pepper

1 teaspoon/3 grams Colman's dry mustard

¼ teaspoon/1 gram ground mace

1½ teaspoons/2 grams grated lemon zest

1½ tablespoons/10 grams chopped fresh flat-leaf parsley

5 feet/2 meters hog casings, soaked in tepid water for at least 30 minutes and rinsed

1. Partially freeze the diced veal and fat, then grind through the large die onto a baking sheet. Return to the freezer until crunchy but not frozen solid, about 20 minutes or so, depending on the temperature of your freezer.

2. Combine the meat and fat with the salt and crushed ice. Regrind this mixture through the small die into the mixer bowl set in a bowl of ice (see Note below).

3. Fit the mixer with the paddle attachment. Add the pepper, mustard, mace, lemon zest, and parsley to the meat mixture and mix on high for 3 to 5 minutes.

4. Do a quenelle test to check the seasoning (keep the rest of the mixture refrigerated while you cook the quenelle), and adjust the seasoning if necessary.

5. Stuff the mixture into the casings and twist into 6-inch/15-centimeter links.

6. Cook the sausages in 170-degree-F./76-degree-C. water to an internal temperature of 150 degrees F./65 degrees C., about 15 to 20 minutes, transfer them to an ice bath and chill thoroughly (to room temperature or below). Refrigerate until ready to use.

7. To serve, gently grill, sauté, or roast until heated through.

Yield: About 2 pounds/1 kilogram sausage;
about eight 6-inch/15-centimeter links

[NOTE: See pages 132–138 for a detailed description of the basic grinding, mixing, stuffing, and cooking techniques.]

MORTADELLA

Mortadella is a mild smooth sausage known for its prodigious girth and the large chunks of white pork fat that stud its interior. It originated in Bologna, Italy, and is similar in appearance and texture to the American sandwich meat popularized by Oscar Mayer, but the similarity ends there. It's traditionally stuffed into beef middles, the big suckers, and either cut into cubes (as is more typical in Italy) or sliced thin, as part of a charcuterie plate. It would certainly be possible to make a large roulade using plastic wrap and poach it (think of it as a gigantic quenelle test–see page 136; it can be tricky, though, because of its size). This is a traditional 5-4-3 forcemeat, but its size–4 to 5 inches/10 to 12.5 centimeters in diameter, sometimes quite a bit larger–and the chunks of fat and other garnish options, such as pistachios, peppercorns, and olives, make it a very special sausage. And that's no baloney.

1 pound/450 grams boneless lean pork shoulder butt, diced
2 tablespoons/30 milliliters dry white wine
½ teaspoon/3 grams minced garlic

1½ tablespoons/20 grams kosher salt

½ teaspoon/3 grams pink salt

12 ounces/335 grams pork back fat, diced

10 ounces/280 grams crushed ice

1½ teaspoons/4 grams ground white pepper

1 teaspoon/4 grams ground mace

1 teaspoon/2 grams ground coriander seeds, toasted (see
    Note page 51)

¼ teaspoon/0.5 gram ground bay leaf

½ teaspoon/1 gram freshly grated nutmeg

½ cup/120 grams finely diced pork back fat, blanched in
    boiling water for 1 minute, drained, and cooled

½ cup/80 grams pistachios, blanched and peeled

1 large beef middle, soaked in tepid water for at least
    30 minutes and thoroughly rinsed (or use plastic wrap)

1. Combine the pork butt with the white wine, garlic, salt, and pink salt. Grind through the small die into a bowl set in ice (see Note below); immediately refrigerate. Grind the fat through the same die into another bowl set in ice; immediately freeze (keep the meat and fat separate).

2. Place the ground meat in the food processor, along with the ice, pepper, mace, coriander, bay leaf, and nutmeg, and process until the mixture reaches 40 degrees F./4 degrees C., about 5 minutes. Add the ground fat and continue to process until the mixture reaches a temperature of 45 degrees F./7 degrees C., about 5 minutes.

3. Place the mixture in a bowl set in ice and fold in the blanched fat and pistachios.

4. Do a quenelle test to check the seasoning (keep the rest of the mixture refrigerated while you cook the quenelle), and adjust the seasoning if necessary.

5. Stuff into the beef middle (or use plastic wrap to form a cylinder 4 to 5 inches/10 to 12.5 centimeters in diameter, and twist and tie the ends to seal).

6. Poach the sausage in 170-degree-F./76-degree-C. water until the internal temperature reaches 150 degrees F./65 degrees C., then transfer to an ice bath to chill completely.

Yield: One 3-pound/1.5 kilogram sausage

[NOTE: See pages 132–138 for a detailed description of the basic grinding, mixing, stuffing, and cooking techniques.]

BOUDIN BLANC

Boudin blanc is a classic French sausage traditionally made at Christmastime. It's rich with whole milk and lots of eggs, which give it a very delicate, mousseline-like texture. *Boudin* means pudding, and this is indeed similar in texture to a pudding. It's flavored with the mixture of spices called *quatre épices,* a combination of pepper, cinnamon, nutmeg, and cloves, a good all-purpose seasoning for pâtés, game dishes, and stews.

It's very important to keep the sausage mixture as cold as possible because it's very loose and as it warms up, it can be difficult to work with, but otherwise it's a sausage that's easily made at home. Brian has served this at the Taste of the NFL Hunger Relief Benefit held during Super Bowl weekend in the host city, a perfect winter offering.

1 pound/450 grams boneless pork shoulder butt, cut into
1-inch/2.5-centimeter dice

1 pound/450 grams skinless, boneless chicken breast, cut into
1-inch/2.5-centimeter dice

1½ tablespoons/20 grams kosher salt

1 teaspoon/3 grams ground white pepper

1½ teaspoons/6 grams Quatre Épices (recipe follows)

8 large eggs

2½ cups/600 milliliters whole milk

3 tablespoons/30 grams all-purpose flour

8 feet/2.5 meters hog casings, soaked in tepid water for at
least 30 minutes and rinsed

1. Grind the meats together into a bowl set in ice (see Note below). Place in the freezer for 5 minutes.

2. Combine the ground meat, salt, pepper, and quatre épices in the processor and process to combine, about 30 seconds. With the processor running, add the eggs one or two at a time, followed by the milk, then the flour. You may need to stop occasionally to scrape down the sides of the bowl.

3. Do a quenelle test to check the seasoning (keep the remaining mixture refrigerated while you do so), and adjust the seasoning if necessary.

4. Stuff the sausage into the hog casings.

5. Poach the sausages in 170-degree-F./76-degree-C. water to an internal temperature of 160 degrees F./71 degrees C. Transfer to an ice bath to chill.

6. Sauté the boudin gently in whole butter over medium-low heat until it's appealingly browned and warmed through.

Yield: About 4 pounds/2 kilograms sausage;
sixteen 6-inch/15-centimeter links

[NOTE: See pages 132–138 for a detailed description of the basic grinding, mixing, stuffing, and cooking techniques.]

Quatre Épices

3 tablespoons/30 grams black peppercorns
1 tablespoon/8 grams freshly grated nutmeg
2 teaspoons/6 grams ground cinnamon
2 teaspoons/6 grams whole cloves

Grind all the spices together in a spice or coffee mill. Store in an airtight container in a cool, dry place.

Yield: ⅓ cup/50 grams

BOUDIN NOIR WITH APPLES AND ONIONS

This is an exciting recipe to do and very easy as long as you have access to pig's blood (which is not so easy to obtain; see below). In fact, it's probably the easiest emulsified sausage to make. The blood is mixed with sautéed onion and apple, then ladled into a funnel to which the casing has been attached. The stuffing is mainly onion and apple; the blood holds it together. You'll need a large funnel to fill the casing, one with an opening big enough for the diced onion and apple to pass through it (1 to 1½ inches/2.5 to 3.5 centimeters in diameter): The color is an extraordinary lavender, very beautiful, until it's poached, when it becomes nearly black–thus its name, black pudding.

This sausage, which has a very rich, savory, but not overly strong flavor, should be eaten as close to making it as possible, that day or the next.

Only pig's blood works here (part of the magic of the pig–even its blood is superior to that of other animals); we don't recommend using beef blood. But as it's illegal to sell pork blood in the United States, it's hard to find. If you know a local grower or the person who dresses the hogs, they may arrange to save some of the blood for you. Some crafty shoppers who live near major Chinatowns may be able to suss out some fresh blood there.

> 2 pounds/900 grams pork back fat, cut into small dice
> 2 pounds/900 grams onions, cut into small dice
> 4 ounces/125 grams unsalted butter
> 3 Granny Smith apples, peeled, cored, and cut into small dice
> ¼ cup/60 milliliters Cognac or Calvados
> 1 cup/250 milliliters heavy cream
> 2 large eggs
> 1 quart/1 liter fresh pig's blood
> 3 tablespoons plus 1 teaspoon/45 grams kosher salt
> 2 teaspoons/6 grams ground white pepper
> 1 teaspoon/4 grams Pâté Spice (recipe follows) or
>   Quatre Épices (page 143)
> 1 tablespoon/6 grams chopped fresh chervil
> 1 tablespoon/6 grams chopped fresh chives
>
> 10 feet/2 meters hog casings, soaked in tepid water for at
>   least 30 minutes, rinsed, cut into 2-foot/60-centimeter
>   lengths, and tied with butcher's string at one end

1. Bring a medium pot of water to a boil. Add the pork fat, reduce the heat to a simmer, and poach for 20 minutes. Drain and set aside.

2. Sauté the onions in 2 ounces/50 grams of the butter in a large sauté pan over medium heat until soft and translucent. Take the pan off the heat, add the fat, and set aside.

3. Sauté the apples in 2 ounces/50 grams of the butter in a large sauté pan over medium-high heat until soft. Turn the heat to high, add the Cognac or Calvados, and cook until it has evaporated. Transfer to a large bowl.

4. Add the onions to the apples and refrigerate till chilled.

5. Combine the cream and eggs in a bowl large enough to hold all the ingredients, and whisk to blend. Strain the blood into the cream (if the blood is congealed, rapidly whisk or blend it with a hand blender to liquefy it). Stir in the remaining seasonings and herbs. Gently stir in the chilled apple mixture. Sauté about ¼ cup/60 milliliters of the mixture, taste, and adjust the seasoning if necessary.

6. Fit a casing over the opening of the funnel and ladle in enough of the mixture to fill it (see Note below). Tie it off with butcher's string and connect the two ends to form a ring. Repeat with the remaining casings.

7. Poach the sausage in 170-degree-F./76-degree-C. water until it's firm and brown juices, not blood, ooze from it when pricked with a needle, 20 to 30 minutes. Remove to a rack to cool, then refrigerate.

8. To reheat, place the sausage on a baking sheet in a 350-degree-F./175-degree-C. oven for about 15 minutes.

Yield: About 5 pounds/2.25 kilograms sausage; 5 large rings

[NOTE: See pages 132–138 for a detailed description of the basic grinding, mixing, stuffing, and cooking techniques.]

Pâté Spice

This is Brian's alternative to the traditional quatre épices mixture often used to season pâtés. Increase or reduce the amounts of the ingredients to suit your own taste and make your pâtés distinctly your own.

> 1 teaspoon/4 grams ground cloves
> 1 teaspoon/4 grams ground nutmeg
> 1 teaspoon/3 grams ground ginger
> 1 teaspoon/3 grams ground coriander
> 2 teaspoons/6 grams ground cinnamon
> 1 tablespoon/10 grams white pepper

Combine all the ingredients and mix well. Store in an airtight container in a cool, dry place in a container with a tight fitting lid.

Yield: 3 tablespoons/30 grams

SHRIMP, LOBSTER, AND LEEK SAUSAGE

This is a fancy seafood sausage using a standard mousseline forcemeat, with large chunks of lobster (a mousseline is a mixture of fish or meat pureed with egg whites and/or cream). It's delicately flavored and works beautifully with a gentle, creamy sauce, such as a lobster sauce or a creamy leek sauce. This kind of forcemeat is one of the easiest and most stable to do at home, and it's a great way to turn a little bit of lobster into a dish that serves many people.

Fines herbes is a stellar all-purpose mixture of the soft herbs tarragon, parsley, chives, and chervil. It's magnificent with veal, chicken, white fish, eggs, steamed new potatoes, and in many other dishes. To make it, simply combine equal parts of the minced herbs. Chervil is a very delicate and shapely herb with an anise flavor that's a little sweeter and less assertive than tarragon; it can be hard to find in supermarkets, though, and can be omitted.

1 pound/450 grams peeled rock shrimp

1 large egg white

1½ cups/375 milliliters heavy cream

2 teaspoons/10 grams kosher salt

¼ teaspoon/1 gram ground white pepper

⅓ cup/50 grams diced leeks, blanched in boiling water until tender and chilled in an ice bath

1 cup/150 grams cooked lobster (about 5 ounces; from a 1¼ pound/625-gram lobster)

1½ tablespoons/10 grams fines herbes (minced fresh flat-leaf parsley, chives, tarragon, and/or chervil, as available)

6 feet/1.5 meters sheep casings, soaked in tepid water for at least 30 minutes and rinsed

1. Puree the rock shrimp with the egg white in a food processor. With the processor running, add the cream slowly, then the salt and white pepper.

2. In a bowl, combine the mixture with the leeks, lobster, and herbs, folding them in gently but evenly.

3. Stuff into sheep casings and twist into 4-inch/10-centimeter links.

4. Poach in 170-degree-F./76-degree-C. stock or water to an internal temperature of

135 degrees F./55 degrees C. (Alternatively, you can form into a cylinder using plastic wrap, twist and tie the ends securely, and poach in water).

Yield: About twenty 4-inch/10-centimeter links

FOIE GRAS AND SWEETBREAD SAUSAGE

Conceived for the Team 2000 Culinary Olympics by Dan Hugelier, Brian has developed this sausage to make use of foie gras scraps. It's written and scaled for restaurant production. We don't really recommend it for the home cook, largely because it would require the purchase of a whole foie gras, and the best thing to do with that is not to grind it up into sausage. However, if you have a lot of foie gras scraps, or you love the idea of it, this is a superlative sausage and the amounts can be reduced proportionately. The foie gras, which is mainly fat, is emulsified into the beef along with the pork fat, adding a rich foie gras flavor to the sausage. Brian's colleague Michael Symon loved the idea of the foie gras sausage, so he started making one with a duck farce and chunks of marinated foie and making another, purely smooth foie gras sausage, which he then smokes ("It's a foie hot dog," he says).

3 tablespoons plus 1 teaspoon/45 grams kosher salt

1 teaspoon/7 grams pink salt

1 teaspoon/8 grams dextrose (or sugar)

2 ½ pounds/1120 grams boneless lean beef, top round or
       another inexpensive cut, cut into small dice

1 pound/450 grams pork back fat, cut into small dice

1 pound/450 grams foie gras scraps

1.5 pounds/675 grams crushed ice (or frozen braising liquid
       reserved from Braised Sweetbreads, below)

¾ ounce/22 grams Colman's dry mustard

½ ounce/15 grams ground white pepper

¼ teaspoon/1 gram garlic powder

1 cup/100 grams chopped fresh herbs, chilled (any
       combination of fines herbes: flat-leaf parsley, chives,
       tarragon, and chervil)

1 cup/250 grams cooked diced bacon (reserve the fat for
       cooking the mushrooms), cooled or chilled

Braised Sweetbreads (recipe follows)

2 pounds/900 grams wild mushrooms, diced and sautéed in
the reserved bacon fat, cooled or chilled

10 feet/3 meters hog casings, soaked in tepid water for at
least 30 minutes and rinsed

1. Combine salt, pink salt, and dextrose, then toss this cure with the beef. Grind through small die (see Note below). Refrigerate. Grind back fat and foie gras scraps and refrigerate. Keep separate from the meat.

2. Place the ground meat in a food processor, put the ice (or frozen braising liquid) on top of the meat, put the mustard, pepper, and garlic powder on top of the ice.

3. Process until the mixture reaches a temperature of 40 degrees F./4 degrees C. Add the fat and foie gras and process to a temperature of 45 degrees F./7 degrees C.

4. Transfer to a bowl and fold in the herbs, bacon, cooled sweetbreads, and cooled mushrooms.

5. Stuff into the hog casings and twist into 4-inch/10-centimeter links.

6. Poach the sausage in milk and stock with aromatics to an interior temperature of 150 degrees F./65 degrees C. Shock in an ice bath till completely chilled.

Yield: About 6 pounds/3 kilograms sausage;
about forty-eight 4-inch/10-centimeter links

[NOTE: See pages 132–138 for a detailed description of the basic grinding, mixing, stuffing, and cooking techniques.]

BRAISED SWEETBREADS

2 pounds/1 kilogram sweetbreads
2 ounces/50 grams unsalted butter
1 cup/140 grams chopped onions
½ cup/70 grams chopped celery
½ cup/70 grams chopped parsnips
1 tablespoon/10 grams black peppercorns
A few branches of fresh thyme
2 bay leaves
1 bottle (750 milliliters) sweet Madeira

1. Soak the sweetbreads in cold heavily salted water (1 cup per gallon/225 grams per 4 liters) for 12 to 24 hours, refrigerated, to remove excess blood.

2. Preheat the oven to 325 degrees F./160 degrees C.

3. In a heavy-bottomed ovenproof sauté pan large enough to contain the sweetbreads snugly in one layer, melt the butter over medium heat. Sauté the vegetables until soft, about 10 minutes. Add the remaining ingredients (except the sweetbreads), bring to a simmer, and simmer gently for another 10 minutes.

4. Meanwhile, remove the sweetbreads from the salted water and rinse.

5. Add the sweetbreads to the simmering liquid, cover, and place the pan in the oven. Braise until the sweetbreads are tender, about 30 minutes.

6. Remove the sweetbreads from the braising liquid (reserve the liquid), allow them to cool, then remove any remaining pieces of fat or thin membrane. Crumble the sweetbreads into small pieces, about the size of a peanut, and refrigerate until ready to use.

7. Strain the braising liquid through a fine-mesh strainer. Allow it to cool to room temperature, then freeze for use in the sausage.

Yield: 2 pounds/1 kilogram sweetbreads

Smoked Sausage

Smoke can raise the craft of a fresh or emulsified sausage to a higher level. Smoked sausages, which may be cooked or uncooked, take on a rich appealing hue and a satisfying depth of flavor. All begin with either the fresh sausage or the emulsified sausage method. A cooked smoked sausage is smoked at a temperature of about 200 degrees F./93 degrees C., similar to the ideal poaching temperature. An uncooked smoked sausage is cold-smoked, smoked at a temperature that won't cook the meat, 100 degrees F./37 degrees C. or lower. The latter are rarely eaten on their own but are almost always used as ingredients in a larger dish (Cajun andouille in a gumbo, for instance).

The sausages in the recipes for cooked smoked sausages here don't have to be smoked—some of them are delicious roasted—and then, obviously, they are considered to be fresh sausages. But they are in this section for a reason—they're great smoked. That also means they're especially tasty on the grill, smoked or fresh, cooked, covered, over indirect heat. The emulsified sausages in the cooked-smoked section are here because they're best smoked, but you can heat them any way you wish—grill, roast, or sauté—whether they're smoked or not. Once they have been cooked, they can be eaten hot or cold.

All of the fresh sausages in the first part of this chapter can be hot-smoked. Chicken Sausage with Basil and Tomatoes (page 123) is particularly excellent smoked. If hot-smoking any of the recipes in the fresh sausage section, it's essential to add 1 teaspoon/6 grams pink salt (sodium nitrite) to the seasoning mixture for each 5 pounds/2.25 kilograms meat and fat to protect against potential botulism poisoning. All smoked sausages and dry-cured sausages require the additional nitrites (see pages 174–176 for a complete discussion of nitrite and nitrate), because sausages held for long periods at temperatures between 40 and 140 degrees F./4 and 60 degrees C. can become corrupted by bacterial growth. All hot-smoked sausages follow the same general smoking method, though times and temperatures may vary.

Basic Smoked Sausage Technique

The salient facts about smoking sausages are these: We smoke sausages for flavor and color, not for preservation (though acids from the smoke that will stick to the skin will prevent molds to some extent); and, in order for the smoke to flavor and color the sausages, it's got to be able to stick to them—thus the need to dry the sausages, to form a *pellicle*, a tacky surface that the smoke compounds will adhere to. Most sausages need to be dried for a couple hours or overnight before being smoked, till their skin feels dry and slightly tacky.

Other matters are mostly common sense. Hang the sausages on smoke sticks to keep them separate, so they aren't touching one another (triangular dowels work best). Areas that touch won't be smoked, and taste and appearance will be affected. Try to be sure that you have an even number of links, or you'll find you've got an extra link hanging where it shouldn't.

The following guidelines refer to hot-smoking, that is, smoking at temperatures of 200 degrees F./93 degrees C. Monitoring the temperature of the interior of the sausages is important. It's helpful to have a probe thermometer attached to a timer/alarm that can be set to go off when the sausage registers 150 degrees F./65 degrees C. (This is handy for all roasted meats.) Otherwise, you will need to check continually with a standard instant-read thermometer, which is inconvenient, allows for error, and results in both loss of juices from piercing the sausage repeatedly and loss of smoke from opening the door.

Try to keep your smoker below 200 degrees F./93 degrees C., a nice gentle heat. Unless you have a temperature-controlled smoker, you'll have to watch the temperature carefully;

heat will usually be struggling to thwart you. But the longer you can keep the temperature down, the better the results will be.

When cooked smoked sausages are taken out of the smoke, they are, like poached sausages, usually chilled in ice water. This stops the cooking and can also prevent shrinkage. Because smoke is so sticky (as you know if you've ever cleaned a smoker), very little flavor is lost in the ice bath.

There are various other particulars with the individual smoked sausages. Some, for example, are taken out of the smoker and left at room temperature to "bloom"–that is, to take on a beautiful rich brown hue). Any such issues will be addressed within the recipes.

KNACKWURST

This is the most straightforward of the smoked sausages–a fresh sausage that's hot-smoked. The only additional ingredient is the pink salt, which protects against the possibility of botulism.

Our friend Marlies Bailey, a native of Germany now living in our heartland, smack in the middle of Oklahoma, and a great fan of sausages, wrote this in response to our question about the name of the sausage: "'Knacken' means to crack, in the literal sense. *Knackwurst* thus means that when you bite into it, it gives you a good crunchy sound, an explosion of flavor and juices. If you see a woman with a nice firm butt, we say she has a *knackiger Asch*. Ha-ha! Does that give you the feel of the word? It's a specific kind of sausage that is always boiled or steamed, never fried, and it's often larger in diameter than your basic sausage."

This version of knackwurst uses the same variety of meats used in a German bratwurst but it has a higher ratio of veal; it also uses paprika in its seasonings, in addition to being smoked. The paprika and black pepper give it a peppery-hot finish. Because the fat content of today's commercial pork shoulder butt is variable, to ensure a juicy succulent sausage, you may want to replace 1 pound/450 grams of the pork with 1 pound/450 grams back fat.

These sausages can be eaten immediately out of the smoke box, or cooled down in an ice bath and refrigerated for a week to ten days or frozen for up to three months. They can be eaten cold or gently reheated and eaten hot.

3 pounds/1350 grams boneless lean veal shoulder, diced

2 pounds/900 grams boneless well-marbled pork shoulder butt

3 tablespoons/40 grams kosher salt

1 teaspoon/7 grams pink salt

1½ tablespoons/15 grams coarsely ground black pepper

1 teaspoon/4 grams ground mace

1 tablespoon/8 grams Hungarian paprika

½ teaspoon/2 grams ground coriander

¼ teaspoon/1 gram ground allspice

1 cup/250 milliliters ice water

10 feet/3 meters hog casings, soaked in tepid water for at least 30 minutes and rinsed

1. Combine all the ingredients except the water and toss until thoroughly mixed. Chill until ready to grind.

2. Grind the mixture through the small die into a bowl set in ice (see Note below).

3. Add the water to the meat mixture and mix with the paddle attachment (or sturdy spoon) until the water is incorporated and the meat has developed a uniform, sticky appearance, about 1 minute on medium speed.

4. Cook a bite-sized portion of the sausage, taste, and adjust the seasoning if necessary.

5. Stuff the sausage into the hog casings and twist into 6-inch/15-centimeter links. Hang on smoke sticks and let dry for 1 to 2 hours at room temperature or in the refrigerator.

6. Hot-smoke the sausages (see page 74) at a temperature of 180 degrees F./82 degrees C to an internal temperature of 150 degrees F./65 degrees C., about 2 hours. Transfer to an ice bath to chill thoroughly, then refrigerate.

Yield: About 5 pounds/2.25 kilograms sausage;
about twenty 6-inch/15-centimeter links

[NOTE: See pages 132–138 for a detailed description of the basic grinding, mixing, stuffing, and cooking techniques.]

HUNTER SAUSAGE (JAGERWURST)

Another sausage with German roots, jagerwurst is a pork sausage with a chunky country-style grain. A hunter or outdoorsman could easily carry this sausage with him for sustenance, the aggressive seasonings of mustard, coriander, garlic, nutmeg, and ginger making it delicious to eat cold. This is an excellent hors d'oeuvre sausage for that same reason, but it's also superb grilled, sautéed, or gently roasted just till warmed through.

The chunky texture is the result of two different grinds. All the meat is passed through the large die, then half of it through the small die. The result is large chunks of meat in distinct shades of pink punctuated here and there by satiny white fat; these qualities are called the sausage's "definition" in the sausage world.

5 pounds/2.25 kilograms boneless pork shoulder butt, diced

3 tablespoons/40 grams kosher salt

1 teaspoon/7 grams pink salt

1½ teaspoons/5 grams coarsely ground black pepper

½ teaspoon/2 grams ground coriander seeds, toasted (see Note page 51)

2 teaspoons/12 grams minced garlic

2 tablespoons/16 grams yellow mustard seeds, toasted (see Note page 51)

1 teaspoon/4 grams freshly grated nutmeg

½ teaspoon/2 grams ground ginger

10 feet/3 meters hog casings, soaked in tepid water for at least 30 minutes and rinsed

1. Combine all the ingredients and toss until thoroughly mixed. Chill until ready to grind.

2. Grind on the large die into a bowl set in ice (see Note below). Grind half of this mixture through the small die.

3. Combine the two ground meats and mix with the paddle attachment (or a sturdy spoon) until the meat appears sticky, 2 to 3 minutes on medium speed.

4. Stuff the sausage into the hog casings and twist into 6-inch/15-centimeter links. Hang on smoke sticks and allow to dry for 1 to 2 hours at room temperature, with good ventilation, or in the refrigerator.

5. Hot-smoke the sausages (see page 74) at a temperature of 180 degrees F./82 degrees C. to an internal temperature of 150 degrees F./65 degrees C. Transfer to an ice bath to chill thoroughly, then refrigerate.

Yield: About 5 pounds/2.25 kilograms sausage;
about twenty 6-inch/15-centimeter links

[NOTE: See pages 132–138 for a detailed description of the basic grinding, mixing, stuffing, and cooking techniques.]

SMOKED ANDOUILLE

This sausage is so flavorful you can eat it plain with no accompaniments. Combining subtle seasonings with some sweet aromatic juiciness from the onion, it's delicious. Smoked andouille is excellent in bean soups and stews, in gumbo, or as a component of a hearty meat stew.

The variations on andouille are many. Andouille can be fatty and is usually heavily spiced and smoked (as this one is), and it may include pig offal (which this doesn't). Andouille is often sliced thin and eaten cold. This one is delicious hot as well. And the sausage can be stuffed into sheep casings for skinny sausages.

Andouille shows up frequently in Cajun cooking; this variety is often cold-smoked. In some parts of France, *andouille* simply refer to any sausage that's cooked and served with beans.

Andouille should not be confused with andouillettes, especially if you are dining at the bouchons of Lyons and have a timid palate. If you order an andouillette, you will be served a large sausage composed entirely of chitterlings–chopped pig intestine–a revered specialty of the region.

5 pounds/2.25 kilograms boneless pork shoulder butt, diced
3 tablespoons/40 grams kosher salt
2 teaspoons/6 grams cayenne pepper
1 teaspoon/7 grams pink salt
1 teaspoon/1 gram dried thyme
½ teaspoon/2 grams ground mace

½ teaspoon/2 grams ground cloves

⅛ teaspoon/1 gram ground allspice

¾ teaspoon/3 grams Colman's dry mustard

1 cup/140 grams diced onion

1 tablespoon/18 grams minced garlic

10 feet/3 meters hog casings, soaked in tepid water for at
least 30 minutes and rinsed

1. Combine all the ingredients and toss to mix thoroughly. Chill until ready to grind.

2. Grind the mixture through the small die into a bowl set in ice (see Note below).

3. Mix with the paddle attachment (or a sturdy spoon) until the meat mixture develops a uniform, sticky appearance, about 1 minute on medium speed.

4. Cook a bite-sized portion of the sausage, taste, and adjust the seasoning if necessary.

5. Stuff the sausage into the hog casings, and twist into 6-inch/15-centimeter links. Hang on smoke sticks and let dry for 1 to 2 hours at room temperature or in the refrigerator to develop the pellicle.

6. Hot-smoke the sausages (see page 74) at a temperature of 180 degrees F./ 82 degrees C. to an internal temperature of 150 degrees F./65 degrees C. Transfer to an ice bath to chill thoroughly, then refrigerate.

Yield: About 5 pounds/2.25 kilograms sausage;
about twenty 6-inch/15-centimeter links

[NOTE: See page 132–138 for a detailed description of the basic grinding, mixing, stuffing, and cooking techniques.]

CHEF MILOS'S COUNTRY VENISON SAUSAGE

Milos Cihelka, Brian's mentor, emigrated from Prague to Canada, where he found work at a lodge that allowed him to ski during the day and cook at night. Venison is big in those rustic climes, and there he created this superb venison sausage that Brian has been making and teaching for twenty years. It's a beautifully seasoned, very rich, red-meat–flavored sausage.

3 ½ pounds/1.5 kilograms boneless lean venison, all fat and
sinew removed, diced

1½ pounds/675 grams boneless pork shoulder butt, diced

2 tablespoons/30 grams sugar

3 tablespoons/40 grams kosher salt

1 tablespoon/6 grams onion powder

1 teaspoon/3 grams ground white pepper

2 teaspoons/6 grams Hungarian paprika

1 teaspoon/7 grams pink salt

½ teaspoon/2 grams ground allspice

½ teaspoon/2 grams freshly grated nutmeg

1 teaspoon/3 grams freshly ground black pepper

½ teaspoon/1.5 grams garlic powder

1 cup/250 milliliters ice water

10 feet/3 meters hog casings, soaked in tepid water for at
least 30 minutes and rinsed

1. Combine all the ingredients except the water and toss to mix thoroughly. Chill until ready to grind.

2. Grind the mixture through the small die into a bowl set in ice (see Note below).

3. Add the water to the meat mixture and mix with the paddle attachment (or a sturdy spoon) until the water is incorporated and the mixture develops a uniform, sticky appearance, about 1 minute on medium speed.

4. Sauté a bite-sized portion of the sausage, taste, and adjust seasoning if necessary.

5. Stuff the sausage into the hog casings, and twist into 6-inch/15-centimeter links. Hang on smoke sticks and let dry for 1 to 2 hours at room temperature or in refrigerator.

6. Hot-smoke the sausages (see page 74) at a temperature of 180 degrees F./82 degrees C. to an internal temperature of 150 degrees F./65 degrees C. Transfer to an ice bath to chill thoroughly, then refrigerate.

Yield: About 5 pounds/2.25 kilograms sausage;
about twenty 6-inch/15-centimeter links

[NOTE: See pages 132–138 for a detailed description of the basic grinding, mixing, stuffing, and cooking techniques.]

SUMMER SAUSAGE

This is a fermented-style sausage, one of the most popular and delicious types of sausage. A fermented sausage is one in which an active bacterial culture feeds on the sugars in the meat mixture, releasing acid, which in turn inhibits the growth of harmful bacteria and creates a pleasingly acidic taste, as with Italian soppressata. It was originally called summer sausage because it kept well in summer's heat without refrigeration. Fermento, a dairy-based flavoring used here, gives the sausage the same tangy flavor, distinct from other acids such as citric acid or a vinegar, but it does not actually ferment. The sausage is ground once and refrigerated to allow it to cure before it is ground a second time; this will result in a firmer-textured sausage and a brighter color. This is excellent sliced thin, served at room temperature with some good mustard. Brian often bakes this sausage inside brioche.

3 pounds/1350 grams boneless lean beef (stew beef,
    chuck roast, round), fat and sinew removed
1½ pounds/675 grams boneless pork shoulder butt, diced
3 tablespoons/40 grams kosher salt
1 ounce/30 grams dextrose (3 tablespoons)
1 teaspoon/7 grams pink salt
½ cup/80 grams Fermento (see Sources, page 303)
4 teaspoons/16 grams Colman's dry mustard
1½ teaspoons/4 grams ground coriander
1 teaspoon/2 grams garlic powder
8 ounces/225 grams pork back fat, diced

10 feet/3 meters hog casings, soaked in tepid water for at
    least 30 minutes and rinsed

1. Combine the beef, pork, salt, dextrose, and pink salt and toss to mix well.
2. Grind the mixture through the large die into a bowl set in ice (see Note below).
3. In a small bowl, dissolve the Fermento in just enough water (¼ to ½ cup/60 to 125 milliliters) to make a thin paste. Add the mustard, coriander, and garlic powder and stir to mix thoroughly. Add to the ground meat mixture and mix with the paddle attachment (or a sturdy spoon) for 2 minutes. Fold in the diced fat.

4. Pack the mixture into a pan or plastic container, pressing out any air pockets. Cover with plastic wrap, pressing down on it so that it touches the meat (no air should touch the meat). Refrigerate for 2 days.

5. Regrind the mixture through the small die. Sauté a bite-sized portion of the sausage, taste, and adjust the seasoning if necessary.

6. Stuff the sausage into the hog casings, and twist into 6-inch/15-centimeter links. Hang on smoke sticks and let dry for 1 to 2 hours at room temperature or in the refrigerator.

7. Cold-smoke the sausages (see page 74) for 2 hours at the lowest possible temperature (to increase its time in the smoke). Turn the heat up to 180 degrees F./82 degrees C. and smoke to an internal temperature of 150 degrees F./65 degrees C. It will have an even brown color and be firm.

8. Remove the sausages from the smoker and hang at room temperature for 2 hours to "bloom" turning the color to a deep mahogany. Refrigerate.

Yield: About 5 pounds/2.25 kilograms sausage;
about twenty 6-inch/15-centimeter links

[NOTE: See pages 132–138 for a detailed description of the basic grinding, mixing, stuffing, and cooking techniques.]

THURINGER

Thuringia is a region in east-central Germany, not far from Leipzig, that is famous for its sausage; this is just one variety. Similar to a summer sausage, it's a fermented (and therefore tangy) sausage, with chunks of fat visible when it's sliced, that can be smoked or not. It can be served hot or cold, sliced, with some good mustard.

4 pounds/1.75 kilograms boneless pork shoulder butt, diced
½ cup/80 grams Fermento (see Sources, page 303)
1 pound/450 grams pork back fat, diced
3 tablespoons/40 grams kosher salt
1 teaspoon/7 grams pink salt

2 tablespoons/20 grams dextrose (or sugar)

2 teaspoons/8 grams black peppercorns, soaked for at least
1 hour in warm water

½ teaspoon/2 grams Colman's dry mustard

2 teaspoons/8 grams ground coriander

10 feet/3 meters hog casings, soaked in tepid water for at
least 30 minutes and rinsed

1. Grind the pork shoulder through the large die into a bowl set in ice (see Note below).

2. Dissolve the Fermento in just enough water (¼ to ½ cup/60 to 125 milliliters) to make a thin paste. Add to the meat, along with all the other ingredients, and mix thoroughly by hand.

3. Pack the mixture into a pan or plastic container, pressing out any air pockets. Cover with plastic wrap, pressing down on it so that it touches the meat (no air should touch the meat). Refrigerate for 3 days.

4. Regrind the mixture through the small die. Sauté a bite-sized portion of the sausage, taste, and adjust the seasoning if necessary.

5. Stuff the sausage into the hog casings, and twist into 6-inch/15-centimeter links. Hang on smoke sticks and let dry for 10 hours at room temperature (65 to 70 degrees F./18 to 21 degrees C.).

6. Cold-smoke the sausages (see page 74) at the lowest-possible temperature, ideally below 100 degrees F./37 degrees C. for 5 hours (the idea is to get as much smoke on the sausages as possible before cooking them in the smoker).

7. Raise the temperature of the smoke box to 180 degrees F./82 degrees C. and smoke the sausages to an internal temperature of 150 degrees F./65 degrees C. Transfer to an ice bath to chill completely, then refrigerate.

Yield: About 5 pounds/2.25 kilograms sausage;
about twenty 6-inch/15-centimeter links

[NOTE: See pages 132–138 for a detailed description of the basic grinding, mixing, stuffing, and cooking techniques.]

SMOKED CHICKEN AND ROASTED GARLIC SAUSAGE

One of our favorite sausages, especially when reheated on the grill. The herbs give it freshness, the garlic an aromatic sweetness. And while it is a chicken sausage, don't think that that means it's low fat; as with every sausage, its deliciousness is proportionate to its fat content.

3½ pounds/1.5 kilograms boneless, skinless chicken thighs, diced

1½ pounds/675 grams boneless pork shoulder butt, diced

3 tablespoons/40 grams kosher salt

1 teaspoon/7 grams pink salt

3 tablespoons/18 grams fresh thyme leaves

3 tablespoons/60 grams Steam-Roasted Garlic Paste (page 125)

1 tablespoon/10 grams coarsely ground black pepper

¼ cup/60 milliliters Chicken Stock (page 226; substitute red wine or water if you don't have homemade stock on hand), chilled

¼ cup/60 milliliters dry red wine, chilled

10 feet/3 meters hog casings, soaked in tepid water for at least 30 minutes and rinsed

1. Combine all the ingredients except the stock and wine and toss to mix thoroughly.

2. Grind the mixture through the small die into a bowl die set in ice (see Note below).

3. Add the stock and wine to the chicken mixture and mix with the paddle attachment (or a sturdy spoon) until the water is incorporated and the mixture develops a uniform, sticky appearance, about 1 minute on medium speed.

4. Sauté a bite-sized portion of the sausage, taste, and adjust the seasoning if necessary.

5. Stuff the sausage into the hog casings, and twist into 6-inch/15-centimeter links. Hang on smoke sticks and let dry for 1 to 2 hours at room temperature or in refrigerator.

6. Hot-smoke the sausages (see page 74) at a temperature of 180 degrees F./82 degrees C. to an internal temperature of 160 degrees F./71 degrees C. Transfer to an ice bath to chill thoroughly, then refrigerate.

Yield: About 5 pounds/2.25 kilograms sausage;
about twenty 6-inch/15-centimeter links

[NOTE: See pages 132–138 for a detailed description of the basic grinding, mixing, stuffing, and cooking techniques.]

KIELBASA (POLISH SMOKED SAUSAGE)

Kielbasa is a generic term for sausage in Poland. Usually it's seasoned forcefully with garlic, and it can be smoked, but the varieties of kielbasa are many: kabanosy, a thin, air-dried sausage flavored with caraway; krakow, thick, hot-smoked, and flavored with garlic and coriander; and wiejska, flavored with marjoram.

This sausage is similar to the kielbasa made from beef that you find in the grocery store—both emulsified and smoked—and very simply seasoned. It's fun to make at home. It's also best not twisted into links, but rather tied into large rings using butcher's string, which is both traditional and visually appealing. It's delicious reheated in Home-Cured Sauerkraut (page 70).

1¼ pounds/565 grams boneless lean beef (stew beef,
    chuck roast, round), fat and sinew removed, diced

1 pound/450 grams pork back fat, diced

4 teaspoons/18 grams kosher salt

½ teaspoon/3 grams pink salt

¾ tablespoon/15 grams dextrose (or 1½ teaspoons/10 grams
    sugar)

12 ounces/335 grams crushed ice

2 teaspoons/6 grams ground white pepper

1½ teaspoons/6 grams Colman's dry mustard

⅛ teaspoon/1 gram garlic powder

9 feet/3 meters hog casings, soaked in tepid water for at least
    30 minutes, rinsed, and cut into six 18-inch/
    45-centimeter lengths

1. Partially freeze the diced beef and fat, then grind through the large die onto a baking sheet (see Note below). Return it to the freezer until it's crunchy but not frozen solid, about 20 minutes.

2. Combine the meat and fat with the salt, pink salt, dextrose, and ice. Grind through the small die into the mixer bowl set in ice.

3. Add the pepper, mustard, and garlic powder and mix with the paddle attachment on high speed for 3 to 5 minutes.

4. Do a quenelle test to check for seasoning (keep the remaining mixture refrigerated while you do so), and adjust the seasoning if necessary.

5. Stuff the mixture into the casings and tie into rings. Dry in the refrigerator overnight.

6. Hang the sausages on smoke sticks and hot-smoke (see page 74) at 180 degrees F./82 degrees C. to an internal temperature of 150 degrees F./65 degrees C. Transfer to an ice bath to chill thoroughly, then refrigerate.

Yield: About 3½ pounds/1.5 kilograms sausage; 6 large rings

[NOTE: See pages 132–138 for a detailed description of the basic grinding, mixing, stuffing, and cooking techniques.]

HOT DOGS

Hot dogs derive their distinctive flavor primarily from garlic, paprika, beef fat, and smoke. This recipe uses the meat from beef short ribs because of its high fat content (buy 4 to 5 pounds with the bone for the 2½ pounds needed here); they should be naturally fatty enough to obviate adding more, and the softness of this fat allows it to be uniformly distributed in the mixture. The ground meat is salted for a day to develop the myosin protein that helps give the hot dog a good bind and a good bite. The mixture is then partially frozen, reground, and rechilled before being pureed to ensure a stable emulsion. Imported in the late nineteenth century by German and Austrian immigrants, the hot dog is now embraced as an American dish. While Chicago is famous for serving them with electric-green relish, mustard, onions, a long pickle slice, sliced tomato, hot peppers, and celery salt, a truly great hot dog–all beef with a natural casing–deserves a simple framework: a plain steamed bun, minced onion, and a good mustard.

2½ pounds/1.25 kilograms beef short rib meat, diced and
    well chilled
1 tablespoon/15 grams kosher salt
1 teaspoon/7 grams pink salt
1 cup/250 milliliters ice water
1 tablespoon/9 grams dry mustard
2 teaspoons/6 grams Hungarian paprika
1 teaspoon/3 grams ground coriander
¼ teaspoon/2 grams ground white pepper
1 tablespoon/18 grams minced garlic
2 tablespoons/30 milliliters light corn syrup

5 feet/1.5 meters hog casings or 10 feet/3 meters sheep
    casings, soaked in tepid water for at least 30 minutes
    and rinsed

1. Grind the meat through a small die (see Note below).

2. Combine the meat with the salt, pink salt, and water and mix by hand to distribute the salts. Cover and refrigerate for 24 to 48 hours.

3. Add the remaining ingredients and mix in by hand, then spread this mixture onto a sheet tray and place it in the freezer until the meat is so cold that it's stiff, about 30 minutes or longer, depending on the freezer.

4. Regrind the mixture through a small die.

5. Return mixture to the sheet tray and place it in the freezer until it is again so cold that it's stiff.

6. Place the mixture in a food processor and puree until it is a uniform paste, about 1 to 2 minutes (if your food processer is not strong enough to do this much meat, simply puree in two batches). It's important not to let the mixture get warm.

7. Stuff the mixture into casings and twist into 6-inch/15-centimeter links. Hang on smoke sticks and hot-smoke (see page 74) to an internal temperature of 140 degrees F./60 degrees C. Transfer to an ice bath to chill thoroughly.

Yield: About 2½ pounds/1 kilogram hot dogs; about ten 6-inch/15-centimeter
links or, if using sheep casings, ten 12-inch/30-centimeter links

[NOTE: See pages 132–138 for a detailed description of the basic grinding, mixing, stuffing, and cooking techniques.]

Basic Uncooked Smoked Sausage Technique

This method is identical to the smoked and cooked sausage technique in all ways except temperature; the sausages are smoked at temperatures too low to cook them. Because they are not fully cooked after smoking, they must be cooked before serving. These sausages are typically not eaten on their own, but rather as a component of another dish, such as a paella.

If you don't have a smoker that can cool the smoke somehow, there are few ways to achieve some level of cold-smoking. If you live in a northern latitude, smoke when it's freezing cold outside, opening the smoker door every so often to let in cold air–it's worth the loss of smoke. At other times and other latitudes, you can put a tray of ice directly above the smoke, which will cool the smoke.

Once these sausages are smoked, they are dried for several days, which helps to intensify their flavor.

They're typically coarse sausages, powerfully flavored.

HUNGARIAN PAPRIKA SAUSAGE

The cold-smoked method results in a dense, flavorful sausage. This particular sausage is all about the paprika–use the best quality you can find. It can be hot or sweet or a combination of both, depending on your tastes, but if you have access to great Hungarian paprika, that's the one to choose for this sausage.

> 1 pound/450 grams boneless lean beef (stew beef,
>   chuck roast, round), fat and sinew removed, diced
> 2 ½ pounds/1 kilogram boneless pork shoulder butt, diced
> 1 pound/450 grams pork back fat, diced
> 3 tablespoons/40 grams kosher salt
> 1 teaspoon/7 grams pink salt
> 1 teaspoon/4 grams ground white pepper
> 5 tablespoons/40 grams best-quality paprika,
>   preferably Hungarian
> ¼ cup/60 milliliters ice water

**10 feet/3 meters hog casings, soaked in tepid water for at
least 30 minutes and rinsed**

1. Grind the pork and beef through the large die in a bowl set in ice (see Note below). Grind the fat through the small die into another bowl set in ice.

2. Combine all the ingredients except the water in the mixer bowl. Mix slowly with the paddle attachment while adding the water. Mix on medium speed until the meat and fat are well distributed and the mixture has developed a sticky appearance, about 2 minutes.

3. Sauté a bite-sized portion of the sausage, taste, and adjust the seasoning if necessary.

4. Stuff the sausage into the hog casings and twist into 10-inch/25-centimeter links. Refrigerate uncovered overnight.

5. Hang the sausages on smoke sticks and cold-smoke (see page 74) for 2 to 4 hours, or until a deep, rich golden brown.

6. Hang the sausages in a cool, dry space (60 degrees F./15 degrees C. is ideal) for 3 days.

7. Refrigerate the sausages for up to a week or freeze until ready to use.

Yield: About 4 pounds/2 kilograms sausage;
about twelve 10-inch/25-centimeter links

[NOTE: See pages 132–138 for a detailed description of the basic grinding, mixing, stuffing, and cooking techniques.]

COLD-SMOKED ANDOUILLE

This is a very spicy sausage, similar in flavor but otherwise quite different from the andouille on page 154. It's traditionally stuffed into sheep casings (though hog casings are all right to use as well) and is not fully cooked until it's used. It's powerful complex flavor makes it too strong to be eaten by itself, but it adds great depth to soups and stews, including traditional Cajun gumbo or jambalaya.

5 pounds/2.25 kilograms boneless pork shoulder butt, diced

1 pound/450 grams yellow or white onions, diced (about
3 cups)

2 tablespoons/36 grams minced garlic

1½ tablespoons/12 grams cayenne pepper

3 tablespoons/40 grams kosher salt

1 teaspoon/7 grams pink salt

½ teaspoon/1 gram dried thyme

¾ teaspoon/2 grams ground mace

⅛ teaspoon/0.5 gram ground cloves

¾ teaspoon/2 grams ground allspice

¾ teaspoon/1 gram dried marjoram

20 feet/6 meters sheep casings, soaked in tepid water for at
least 30 minutes and rinsed

1. Combine all the ingredients and grind through the small die into a mixing bowl set in ice (see Note below).

2. Mix on medium speed with the paddle attachment. Slowly add water and mix until the mixture has developed a sticky appearance, about 2 minutes.

3. Sauté a small portion and taste to check for seasoning.

4. Stuff the sausage into the sheep casings and twist into 10-inch/25-centimeter links. Refrigerate uncovered overnight.

5. Hang the sausages on smoke sticks and cold-smoke (see page 74) for 2 to 4 hours, or until golden brown.

6. Hang the sausages in a cool, dry space (60 degrees F./15 degrees C. with 65 percent humidity is ideal) for 2 to 3 days.

7. Refrigerate the sausages for up to 2 weeks or freeze until ready to use.

Yield: About 6 pounds/2.75 kilograms sausage;
about thirty 10-inch/25-centimeter sausages

[NOTE: See pages 132–138 for a detailed description of the basic grinding, mixing, stuffing, and cooking techniques.]

COLD-SMOKED CHORIZO

Again it's the assertive seasoning that makes this a great sausage. This chorizo is not intended to be especially hot, but you could easily raise the heat level by replacing half a tablespoon of the ancho chile powder with dried chipotle powder or, if you prefer chipotles in adobo sauce, simply add a few minced chipotles. This is an excellent sausage to use in a standard rice pilaf: Sauté minced onion, add the rice, then add the stock, bring to a simmer, and drop in the sausages. Cover the pan and pop it into a 350-degree-F./175-degree-C. oven for about twenty minutes. The rice absorbs the juices and seasonings from the sausage–no better way to cook it.

3½ pounds/1.5 kilograms boneless lean pork shoulder butt, diced

3 tablespoons/40 grams kosher salt

1 teaspoon/7 grams pink salt

1 teaspoon/3 grams ground white pepper

2 tablespoons/16 grams ground cumin

3 tablespoons/24 grams ancho chile powder (or other pure chile powder)

1½ pounds/675 grams pork back fat, diced into 1-inch pieces

¼ cup/60 milliliters ice water

¾ cup/100 grams thinly sliced scallions whites and half the green parts

10 feet/3 meters hog casings, soaked in tepid water for at least 30 minutes and rinsed

1. Combine the pork, salt, pink salt, white pepper, cumin, and ancho powder. Grind through the large die into a bowl set in ice (see Note below). Grind the fat through the small die into another bowl set in ice.

2. Combine the pork mixture and fat in the mixer bowl. Mix with the paddle attachment on low speed while slowly adding the water. Add the scallions and mix on medium speed until the water is absorbed and the meat appears sticky, about 2 minutes.

3. Sauté a bite-sized portion, taste, and adjust the seasoning if necessary.

4. Stuff the sausage into the hog casings, and twist into 8-inch/20-centimeter links.

5. Refrigerate uncovered overnight.

6. Cold-smoke the sausages (see page 74) for 2 to 4 hours, or until the desired color.

7. Hang the sausages in a cool, dry space (60 degrees F./15 degrees C. with 65 percent humidity is ideal) 3 to 5 days.

8. Refrigerate the sausages for up to 2 weeks or freeze until ready to use.

Yield: About 5 pounds/2.25 sausage;
about fifteen 8-inch/20-centimeter links

[NOTE: See pages 132–138 for a detailed description of the basic grinding, mixing, stuffing, and cooking techniques.]

5

THE ARTIST AND THE SAUSAGE:
TECHNIQUES AND RECIPES FOR INDIVIDUALISTIC, IDIOSYNCRATIC, AND TEMPERAMENTAL DRY-CURED MEATS

. . . .

Dry-cured sausages and meats are the quintessence of the charcutier's art and the most difficult to achieve because they rely so heavily on the ambient air and humidity. They can drive even professional charcutiers crazy. The recipes in this chapter are the most difficult in the book, but also the most exciting to attempt, and when you nail it, there may be no more satisfying accomplishment in the kitchen.

By far the most interesting and challenging sausage a craftsman can attempt is one that's never cooked at all: a fresh sausage hung to dry, then sliced thin and eaten as is. We know dried sausage as salami, traditional offerings of the Italian *salumeria*, and as the *saucisson sec* hanging in the window of the French charcuterie.

Many of these sausages are fermented (not unlike pickles, or bread leavened by a sourdough starter) by beneficent bacteria feeding off the sugar, generating lactic acid, which protects the meat from spoilage as well as introduces a pleasingly tangy flavor. The dry-cure method also works for various meats—hams, loins and shoulder, jowl, and pure back fat from the pig, as well as cuts of beef and lamb, giving us prosciutto, coppa, guanciale, lardo, bresaola, and lamb prosciutto.

These dry-cured meats and sausages, almost always sliced thin, are dense and chewy, with a strong, dry-cured flavor and smooth, satiny fat. When we eat them, we're most often eating pork that's never gone above room temperature, let alone come close to the 145 degrees F./63 degrees C. recommended by the government. And yet, properly prepared, these are perfectly safe to eat. There really is nothing similar to eating cured raw meat—it has a flavor and an effect like no other food. It's almost as if your body knows it's raw and forbids it but your brain knows it's excellent and safe and wants it, so that danger and desire commingle in the chewing.

Once again, salt is the key component in rendering raw meat safe and edible. Whole cuts, such as hams and hog jowls, are salted, often with added aromatic seasonings, and left to cure for days or weeks. The salt substantially reduces the microbe activity that would otherwise cause rancidity and rot, the drying finishes off the dehydration process.

All dry-cured sausages must include a curing salt, sodium nitrate (such as DQ Curing Salt #2, different from pink salt), to prevent the possibility of botulism contamination in sausages that are stored at above refrigeration temperatures for long periods. The nitrate over time turns into nitrite, which then does the curing work. These sausages may contain pungent spices, or they may be seasoned with just a hint of garlic and pepper. They may be smoked, or they may simply be hung to dry, perhaps for a couple weeks, perhaps for a year. But, as ever, the three basic foundations of this kind of preparation are shared by all: Some form of salt cure reduces the bacterial activity that would cause spoilage; sodium nitrate prevents botulism spores and other harmful bacteria from growing within the meat, as well as flavoring it and keeping it appealingly red; and the sausages or meats are hung to dry. It should be noted that some European sausage makers say they don't use any curing salt at all. We can't recommend this method for dry-cured sausages because we think it's important to eliminate all possibility of botulism contamination (the word *botulism* derives from the Latin word for sausage for a reason).

Humidity is necessary for proper drying. If the air is too dry, the surface of the meat can dry out, harden, and then prevent the interior moisture from escaping, resulting in a rotten sausage. The final component in a sausage is beneficial bacteria, whether present on the meat or added to the mixture (see page 176). These bacteria feed on sugars and release acid as a byproduct, creating an inhospitable environment for bad bacteria and acidity that gives great dry-cured sausages a pleasing tanginess.

Dry sausages begin no differently from fresh sausages. Meat, salt, and seasonings are ground together and stuffed into casings. But instead of being heated, they are dehydrated. Dry-curing results in a beautiful type of sausage, the most individualistic, idiosyncratic, and temperamental sausage there is, precisely because of its reliance on atmospheric conditions, which change all year round, and the presence of varying microflora in the air.

Because these are so temperamental and idiosyncratic, they can also be the most frustrating sausage to attempt. Some friends and I visited Georges Reynon, one of the most respected charcutiers in Lyon, France, a city famed for its dried sausages. Reynon gave us a tour of his shop, where we watched his butchers trimming very thick, exquisitely white back fat of all glands and nerves and impurities (the keys to great dried sausage, Reynon said, are beautiful fat and skillful butchers), and a tour of his drying room. There, where more than a thousand sausages hung, the humidity and temperature were perfectly controlled. Still, Reynon expressed his frustration at being unable to achieve a consistent product. The sausages were never the same, he said, and it drove him crazy. He pointed to some that were perfect, and others that had a thick coating of white mold on them, not harmful but something he considered to be an imperfection. He had no idea why they should be different, but they were.

The process is elusive. You cannot see the water leaving the sausage, but you can feel the sausage becoming firmer and denser, and you can weigh it to gauge the water loss. You can't see the bacteria generating acid, and you'd need a pH meter to know how much acid was actually in the sausage. What is it that is attaching itself to the skin? Hard to say. Smooth white stuff tends to be good, fuzzy is definitely bad, as are any colors other than white. And what's going on inside the sausage? Impossible to say until you cut it open. Dry-curing sausage takes time and patience, experimentation, and a willingness to get it wrong at first.

But when your sausage has dried just right, and you slice it thin, and the interior is a glistening deep crimson red with bright pearly chunks of fat, it is incredibly exciting. This is real mastery over the food we prepare. To make a home-cured pork sausage, with just salt and pepper for seasoning, is a deeply gratifying experience, like making a great wine. Mastering the technique of transforming raw meat and fat, whether a sausage or a whole muscle, into something delicious without using heat, enhances your ability to work with all food. This is true craftsmanship, craftsmanship aiming for art, reliant on the cook's skill and knowledge and, perhaps, a little bit of divine intervention.

While dry-curing is unpredictable and temperamental, and the cook first attempting dry-curing should prepare for early failure, the process is actually very simple. Practically the only special equipment you need is a place to hang your sausage or ham or belly or fat in.

Humidity is the critical factor and the one most difficult to control. Air with less than 60 percent humidity tends to dry out the casings before the water inside can get out, causing rot; 70 percent humidity is better. Cool is also best, about 60 degrees F./15 degrees C. Some circulation is important. And the sausages should be protected from light, which can damage the fat.

How you achieved this may be up to your imagination. An unused refrigerator can be a perfect drying box, given a pan of heavily salted water to create the humidity (the salt keeps molds from growing in the water) during the early stages of fermenting and curing. It shouldn't matter how you create the enclosed humid environment–whether in a room in your cellar or in an old wooden box–but without humidity, you won't be able to dry-cure properly.

With the appropriate humidity, you can dry-cure just about anything. A round of beef becomes bresaola. Chunks of diced pork shoulder stuffed into casing and dry-cured become coppa. Even pure fat can be cured, in the great tradition of lardo, cured back fat. Or, what may be even better in America, given the sorry state of the hog and its diminishing flavors and fat, dry-cured pork belly. Sliced thin and eaten on a crouton (or just straight off the knife), it's a revelation. So much so that several years after *Charcuterie* was originally published, we wrote a book devoted solely to the Italian craft of dry curing meats, called *Salumi*.

The following recipes offer a range of dry-cured preparations. Some are for whole pieces of meat, others call for ground meat and fat. In some recipes, chunks of meat are cured and forced into a casing. Some of these are cold-smoked before being dried, some are hot-smoked first. Some are heavily seasoned, others call for almost no seasoning. You can dry just about any cut of beef or pork with good results provided that it's cured with salt and a curing salt first.

For novices, the best strategy is to start with skinny items, sausages that don't take long to dry, sausage stuffed into sheep casings or hog casings. The longer something takes to dry, the greater potential for problems.

The recipes that follow represent one of humankind's oldest form of food preparation and preservation, one once almost unheard of in the American home kitchen. They represent a craft that's been practiced vigorously throughout Europe for centuries and continues to thrive; here, the craft is in its infancy. All of which makes the dry-cure technique one of the most interesting in the specialty of charcuterie.

Dry-Cure Essentials and Safety Issues

The first essential is great meat. This is especially so for whole items such as hams, lardo, and cured pork belly. Do not use commercially raised pork for these preparations. We recommend using only farm-raised hogs, animals raised out of doors by sustainable farms. There are good mail-order sources and sources on the Internet for excellent pork; see Sources, page 304.

The main special equipment you need to dry-cure meat and sausage is a cool, dark humid space. Some people have a damp cellar with what are optimum conditions, 60 degrees F./15 degrees C. and 70 percent humidity. For those who don't, a second refrigerator, unplugged, makes a good drying box. To create humidity, put several cups of salt in a nonreactive container and fill it with an equal amount of water (the salt will keep mold and bacteria from growing in the water; the water will not likely need to be replaced during drying). One last element: The place in which you dry should be dark. Fat reacts to light—light actually breaks it up—and can become rancid if exposed to too much light.

Safety issues are paramount. Don't play fast and loose with the following techniques, or you could make yourself or someone else sick. Here are the main issues.

SANITATION

Sanitation is always important when preparing food, but it's especially important in dry-curing, because a little bacteria in a sausage, for example, held at room temperature for several weeks can grow into a big problem. But you will not have a problem if you work cleanly, wipe down and sanitize counters and boards, keep knives in their place, and avoid any possible cross-contamination. (This is especially important in restaurant kitchens. We can't recommend strongly enough that chefs dry-curing their own sausages designate a special area for making sausages. Given the range of foods within, and the busyness of, a restaurant kitchen, the possibility of cross-contamination is high.)

Be sure your equipment is clean. Use a bleach solution to sanitize your grinder, stuffer, work surface, and sponge (often the most germy item in a kitchen). We recommend one capful of bleach for each quart of water for an effective sanitizing solution, which will kill salmonella, E. coli, staph, and other bad guys.

To reiterate: Because of the warm, moist conditions involved with dry-curing foods, bacteria can multiply with abandon, so sanitation is a much more salient concern than when working with food that will be cooked or eaten immediately.

SALT

Use it in the quantities indicated in our recipes. Generally speaking, a strong salt concentration is the key factor in preventing anything harmful from growing. We often use more salt in dry-cured sausages than in fresh sausages (our standard ratio is 2.75 percent salt).

CURING SALTS (SODIUM NITRITE AND SODIUM NITRATE)

Some form of curing salt must be used in any dry-cured sausages. The warm, anaerobic, protein-rich interior of a sausage is an ideal environment for the bacteria that produce the potentially fatal nerve toxin causing botulism poisoning. Sodium nitrite, refered to in our recipes as pink salt (see Sources, page 303), prevents these bacteria from growing. Sodium nitrate, which is kind of a time-release form of sodium nitrite, must be used in all dry-cured sausages cured over longer periods. Always use the curing salts in the proportions indicated in the recipes here (our standard ratio is .25 percent for a total salt concentration of 3 percent). Remember that curing salts are dangerous if ingested accidentally, so use exactly as stated in the recipe and keep them well marked.

Nitrites are added to cures for either dry-curing or smoking, processes in which the meat is held at between 40 and 140 degrees F./4 and 60 degrees C., temperatures at which harmful bacteria multiply rapidly. Nitrites have three functions, one related to health, one to appearance, one to flavor: (1) Nitrites kill a range of bacteria, most notably the one responsible for botulism. (2) Nitrites help preserve the pink color that we associate with tasty meats and retard rancid flavors in the fat. A sausage or terrine cured with nitrite will retain a rosy interior hue, one without will turn gray, even when they both have reached an internal temperature of 150 degrees F./65 degrees C. (3) Nitrites add a piquant flavor to meats (a chicken cured with pink salt will have a distinct ham flavor and rosy hue).

Nitrates are also an essential curing agent, but nitrates do nothing beneficial to food until they convert to nitrite. Nitrates act like a time-release capsule for items requiring especially long cures (such as salami, which may be cold-smoked and then dried for weeks). Nitrates began to be added to cured and smoked meats as early as the 1500s in the form of saltpeter, potassium nitrate. The compound was used in America until the 1970s, when it was deemed too inconsistent to be safe (though it remains common in Europe). Sodium nitrate is now manufactured and sold under the names DQ Curing Salt #2 and Insta Cure #2 (see Sources, page 303).

HOW REAL IS THE DANGER OF BOTULISM?

If you get it, very; the toxin is considered to be one of the most poisonous substances on earth.

How likely are you to get it? Not very.

According to the Centers for Disease Control and Prevention, twenty-five or so people a year contract food-borne botulism in the United States, usually as a result of improper home canning. If botulism spores, which are hard to kill, get into a can or jar or are ground into the center of a sausage, then sealed in an oxygen-free, nonacidic environment, they thrive. Botulism has been caused here by canned tuna and garlic stored in olive oil. (In Japan, preserved fish is the chief culprit.)

There are three parts to the botulism equation: Botulism spores develop into the bacteria *Clostridium botulinum*, which in turn produces the deadly toxin. The spores, which are typically found in soil, are difficult to kill but aren't harmful (except, potentially, to infants), and the bacteria themselves aren't harmful either. But allow the bacteria to grow in an anaerobic nonacidic environment between 40 degrees F./4 degrees C. and 140 degrees F./60 degrees C., and they will start producing the deadly toxin.

NITRITES AND YOUR HEALTH: Nitrites remain controversial because they can produce, under certain circumstances, organic compounds called nitrosamines, some of which can cause cancer. Nitrites are found naturally in many vegetables that grow in the ground (add chopped spinach to a white sausage and you may see the meat begin to turn pink, as the nitrites in the spinach begin to cure the meat). We probably consume far fewer nitrites than our ancestors did (at least those who relied far more on preserved food). There is also growing evidence that vitamin C (ascorbic acid) enhances the effect of nitrites while prohibiting the development of nitrosamines in cured and cooked foods such as bacon and some evidence that they're beneficial. At the very least, the evidence suggests that, in limited quantities, nitrites are not a substantial health concern. For a thorough discussion of the controversy, see the redoubtable Harold McGee's *On Food and Cooking*, where he says, "Nitrosamines are known to be powerful DNA-damaging chemicals, yet at present there's no clear evidence that the nitrites in cured meats increase the risk of developing cancer. Still it's probably prudent to eat cured meats in moderation and cook them gently."

When pink salt (sodium nitrite) or the #2 curing salts (sodium nitrate) is listed in one of our recipes, it is critical to use it as directed in order to prevent the potential of botulism poisoning. It is only required for foods that will be smoked or sausages that will be air-dried. Some of the recipes in the book list pink salt as optional; this is because in some preparations, pâtés and terrines for instance, the pink salt adds flavor and a rosy hue that is more appealing than the gray it would otherwise be. But provided these foods are properly refrigerated, the pink salt is not required for food safety issues.

LACTIC ACID AND LIVE CULTURES

Acid is another important part of the dry-cure process. Brian and I recommend using a live culture, called Bactoferm F-RM-52 (see Sources, page 303); it feeds on the sugars in the sausage mixture and releases lactic acid, reducing the pH level and thereby preventing bacterial growth (pH is a measure, scaled from 0 to 14, of the acidity or alkalinity of a given environment). Strips of paper measuring pH are inexpensive and easily available by mail (see Sources, page 306). If you get involved in a lot of dry-curing, it's not a bad idea to invest in a pH meter, at about $350 (see Sources, page 306). Sausage with a pH no higher than 4.9 is considered to be sufficiently acidic to prevent the growth of harmful bacteria.

When working with these live cultures, which are freeze-dried and need to be rehydrated, it's important to use distilled water. Some water is chlorinated, which could kill the live cultures. Please note that while 1 ounce/25 grams of Bactoferm is enough for 220 pounds of sausage, recipes for small batches must contain at least a quarter of the package, about a tablespoon, to ensure that enough of the live culture gets into the sausage; the product is completely safe, adding too much is not harmful in any way.

SUGAR

These recipes call for a sugar called dextrose, which is finely textured glucose derived from cornstarch. It is used in dry-cured sausages because it dissolves more easily and is distributed thoughout the mixture more uniformly than granulated sugar. (See Sources, page 303.)

MOLD

Some mold is desirable, and some mold is harmful. Rule of thumb: Fuzzy mold, no matter what color, is bad, as is any mold that is not white, and should be wiped off the surface with a clean cloth soaked in vinegar or a brine (¼ cup per quart/50 grams per liter is sufficient), just as cheesemakers do with some cheeses. Fuzzy mold (usually it has a greenish cast) can even dig through the casing and damage the interior; if you find evidence of it, to be cautious, throw the sausage away and try again.

Dry white mold, the kind one often sees on dry-cured sausages, is generally considered to be good mold in that it prevents bad molds from growing, feeds on the oxygen at the surface of the sausage, and creates a protective layer, not unlike smoke.

Since this book was originally published, a commercial mold culture, Mold 600, sold by Butcher & Packer (see Sources, page 303), has become available and we recommend it. By actively applying a live mold culture to your sausage you beat the bad guys to the punch and make it difficult for them to grow (and again, it helps protect the sausage in other ways as well). Applying Mold 600 can be considered an optional step for all dry-cured sausages.

PREVENTING TRICHINOSIS BY FREEZING

Trichinosis, a food-borne sickness caused by the larvae of the *Trichinella* worm in pork and wild game, was once common in the United States, mainly contracted by eating pork that hadn't been thoroughly cooked. Today, pork is far less likely to carry the larvae than are wild game, and the disease is relatively rare. About thirty-eight cases were recorded each year during the 1990s, according to the Centers for Disease Control and Prevention. And since then, regulations in how pigs are fed, as well as increasingly informed consumers and the ease of freezing meat, have also contributed to the reduced incidence.

Nevertheless, trichinosis does exist, but preventing the remote possibility of its occurrence is easy, and in some cases a necessary precaution. Though many chefs who dry-cure sausage consider freezing meat sacrilege, as a precaution, we must recommend that pork that is to be dry-cured (that is, not cooked) be frozen before using, especially if it has been raised in a free-range environment. The Centers for Disease Control says that pork less than 6 inches/15 centimeters thick can be frozen for twenty days at 5 degrees F./−15 degrees C. or less to kill the *Trichinella* larva. The freezing times can be shortened by lowering the temperature to −10 degrees F./−23 degrees C. (for 12 days) and −20 degrees F./−30 degrees C. (for 6 days).

CASINGS

Certain dried sausages traditionally use specific casings that determine the diameter of the sausage. Cold-smoked andouille, for example, is prepared in sheep casings, and salami is stuffed into beef middles, a section of cow intestine that's about 3 inches/7.5 centimeters in diameter. While we encourage you to maintain such traditions, from a purely practical standpoint, the casing is not critical. The diameter may change your experience eating it, but it doesn't affect the principles of dry-curing. So you can put any of these sausages into whatever casings are available to you. However, manufactured collagen casings, available in varying sizes, do work, but they're not as satisfying to work with as natural casings.

TROUBLESHOOTING DURING THE DRYING PROCESS

The biggest problem with home-dried sausages is what's called case hardening: a drying out of the casing that prevents water from leaving the sausage. A sausage with case hardening will develop a gray oxidized ring around the outside with a red, not-quite-dried interior. If your humidity is variable, you should examine your sausages daily to check the texture. The casing should feel hydrated, not dried out; it should have a soft texture, some traction when you run your thumb over it. If it feels smooth and dry, and is almost shiny, it is too dry. If you sense the casings are drying out, try misting them with water once a day for the first week.

The second biggest problem is mold. If you're having problems with the green fuzzies, wipe down your drying box with a bleach solution. Wipe down your sausages with brine or vinegar, as directed on page 177. Some experts have recommended hanging a sausage with good mold (see page 177) on it in your drying box: Molds compete, and the more good mold you have, the less likely bad molds are to take over. Or be aggressive and apply a mold culture yourself (see page 176).

The third problem might be achieving a high-enough acidity within the sausage (or a low pH, 4.9 or below). If your sausages are not drying properly and you can't figure out why, they may not have fermented properly. The only way to accurately measure the pH is with pH paper or a pH meter (see Sources, page 306). If it doesn't have a pH of 4.9 or below, odds are it won't dry properly.

DRYING TIMES AND JUDGING DONENESS

Once you've cured your meat, stuffed it into casings, and hung it in your dark, humid, cool place, your next task is to know when it's done. This is difficult to gauge without experience because here doneness isn't measured by temperature; it's determined by water loss.

One way to gauge water loss is to weigh your sausages after you've made them, and make a note of the weight. Then, when you sense that they may be done, weigh them. Generally speaking, a 30 percent weight reduction indicates doneness. Five pounds/2.25 kilograms of fresh sausage should, when dried, weigh about 3 ½ pounds/1.60 kilograms.

Touch is a fairly reliable means of judging the doneness of sausage. Squeeze the sausage: It should feel stiff, almost hard, all the way through to the center. Ultimately, you probably won't know for certain that a sausage has finished drying until you've made a few and acquired a sense of how your drying environment works. Cutting into the sausage is the only sure way of knowing. So, slice off the end of the sausage. If the center still has a raw appearance, if it's got a tacky, squishy texture, return it to the drying box.

The same general rules apply for the whole muscles. The item, whether a ham or a cured loin or an eye of beef should feel dense. When you slice it, the center of the meat should be uniformly dense as well; it shouldn't look or feel raw.

We've given approximate times in the recipes, but these can vary considerably depending on your conditions. It's best to rely on sight and touch. Use your common sense. This may seem obvious, but Brian and I can't overstate it. A finished dry-cured sausage should be completely firm and uniform in color, and it should not look at all raw. Does it look good? Does it look like the dried sausages or sliced salami you've had before? If the casing is hard and the interior still has a mushy raw texture, it's not dried. Does it have a dark ring around the outside of a slice, another indication that the outside hardened before the interior dried? Does the interior smell bad? Any signs of deterioration or rot or rancidity mean that something went wrong: Throw it out. This is one of the few instances in cooking in which mistakes are not necessarily edible. If "mistakes" are cooked to 155 degrees F./68 degrees C. or above, they will be safe to eat, but the flavor may be off. If you've got a bad sausage, throw it out.

Obviously Brian and I can't be in your kitchen making sure your grinder is immaculately clean, that you're adding the right amount of salt and nitrate, and so forth, and so cannot be responsible for you if you get sick. If that concerns you, if you don't feel comfortable evaluating dry-cured meats, don't use dry-cure techniques. (See Sources, pages 304–305, for excellent dry-cured sausages.)

On the other hand, if you follow the directions precisely, act to prevent the potential dangers noted above, and use your common sense, you shouldn't have any safety problems. These techniques have been used at home for centuries.

STORING

The easy part. Keep dry-cured sausages in your refrigerator, preferably wrapped in parchment paper or butcher's paper. They will keep for a month or more if you store them with care.

QUANTITIES IN THESE RECIPES

Most of these recipes call for 5 pounds/2.25 kilograms of meat and fat. As with the other sausage recipes in this book, they can be halved. Remember, though, that the drying process will reduce their weight by about 30 percent–so that 5 pounds/2.25 kilograms of sausage ultimately yields 3½ pounds/1.60 kilograms–and also that they keep for a long time.

THE FIRST TIME

As mentioned earlier, some sausages have specific shapes and sizes, from skinny to fat. As a rule, though, any ground sausage can be stuffed into any kind of casing. Because it is easier to succeed with a thinner sausage and a shorter drying time (meaning less time for case hardening and for bad molds to develop), you might consider beginning with sheep casings or hog casings rather than casings with larger diameters. Any of the sausage recipes will work just as well, and maybe better, in a skinny casing rather than the traditional casings suggested in the recipes. Or you might start with the Landjager on page 191, a German cold-smoked dried sausage, which beginners seem to have a lot of success with. For all long-cured items, don't forget to note on a calendar or in a notebook, the day you begin to cure a sausage and the expected time of completion.

TUSCAN SALAMI

Here is a classical salami with the additional flavors of red wine and fennel; use a good big red wine and fresh fennel seeds. This, as do all the sausages in this chapter, relies on the fermenting process. A beneficial bacteria, sold in the United States as Bactoferm F-RM-52, is introduced to the raw ingredients. It feeds on the sugar and produces lactic acid as a result. This acid gives the salami a pleasing tartness and, more important, the acid makes it difficult for harmful bacteria to grow.

The method for this sausage can be used for any of the ground dried sausages. Most types of casings can be used as well, though different diameters will result in differing drying times. Sausages in sheep casings will take about a week to dry (and are thus the easiest); sausages in hog middles, the largest and surprisingly delicate, will take a month or more. But the most easily available and therefore the most practical casing to use here is the hog casing.

One attribute of a great dried sausage is its "definition," which refers to the grind and the contrast of meat and fat in the finished sausage. The fat should be distinct in a dried sausage, which is why the fat and meat are ground separately from each other; often the fat is ground through the large die. If using the large hog middles, you could cut your fat into small dice rather than grind it; we recommend grinding it for the standard hog casings.

1 pound/450 grams pork back fat, diced (see Note 1 below)

4 pounds/1800 grams boneless pork shoulder butt, diced

¼ cup/56 grams kosher salt

1 teaspoon/7 grams DQ Curing Salt #2 or Insta Cure #2
    (see page 105)

1 tablespoon/10 grams Bactoferm F-RM-52 (live starter culture;
    see Sources, page 303)

¼ cup/60 milliliters distilled water

3 tablespoons/30 grams dextrose

1½ tablespoons/12 grams fennel seeds, toasted and cracked
    beneath a small heavy pan or side of a knife

4 teaspoons/12 grams coarsely ground black pepper

1 teaspoon/6 grams minced garlic

4 ounces/125 milliliters Chianti or other dry Tuscan red wine

10 feet/3 meters hog casings, 20 feet/6 meters sheep casings,
    or one 3-foot/1-meter hog middles, soaked in tepid
    water for at least 30 minutes and rinsed

1. If you are not going to use whole chunks of fat, grind the fat while it is still very cold, or even partially frozen, through the large die into a bowl set in ice (see Note 2 below).

2. While the pork is still very cold, or even partially frozen, combine it with the salt and DQ Curing Salt #2 or Insta Cure #2 and grind through the small die into a bowl set in ice.

3. Combine the fat and meat in the bowl of a standing mixer and refrigerate it while you ready the culture and the remaining ingredients. (The purpose of keeping the meat and fat as cold as possible is to keep the fat distinct from the meat in the finished sausage.)

4. Dissolve the Bactoferm in the distilled water and add it to the meat, then add the remaining ingredients. With the paddle attachment, mix until all the ingredients are well distributed, about 1 minute.

5. Stuff the sausage into the casings. If using hog or sheep casings, twist into 8-inch/20-centimeter links. Or, if using a hog middle, tie off into 12-inch/30-centimeter sticks. Using a sterile pin or needle, poke holes all over the casings to remove any air pockets and facilitate drying.

6. Cover the sausage with a clean towel and leave out at room temperature, ideally 85 degrees F./30 degrees C., for 12 hours to "incubate" the bacteria; the beneficial bacteria will grow and produce more lactic acid at a warmer temperature.

7. Hang the sausage (ideally at 60 degrees F./15 degrees C. with 60 to 70 percent humidity) until it is completely stiff throughout and/or it has lost 30 percent of its weight (about 1½ pounds/675 grams), 6 to 8 days if using sheep casings, 12 to 18 days if using hog casings, 25 to 30 days if using hog middles.

Yield: About 3 pounds/1½ kilograms sausage

[NOTES: 1. If you choose to freeze your pork and fat, do so 2 to 3 weeks before making the sausage, according to the instructions on pages 177–178. Thaw the meat in the refrigerator for 1 to 2 days. 2. See pages 105–115 for a detailed description of the basic grinding, mixing, stuffing, and cooking techniques.]

PEPERONE

This heavily seasoned sausage, dating to Roman times, is widely produced in America (where we spell it pepperoni), by virtue of its importance on pizza. Because of this mass production, the version most of us know is a pale imitation of the original peperone. True peperone (the name means large pepper, or large strong-tasting fruit) is a very lean, tangy, highly spiced sausage. Beef is typically used, but pork is also a good meat to use here.

5 pounds/2.25 kilograms boneless lean beef (stew beef,
   chuck roast, round), fat and sinew removed, diced

¼ cup/56 grams kosher salt

1 teaspoon/7 grams DQ Curing Salt #2 or Insta Cure #2 (see
   page 105)

1 tablespoon/10 grams Bactoferm F-RM-52 (live starter culture;
   see Sources, page 303)

¼ cup/60 milliliters distilled water

3 teaspoons/9 grams cayenne pepper

½ teaspoon/1 gram ground allspice

1 teaspoon/2 grams ground fennel

4 tablespoons/40 grams dextrose

2 tablespoons/16 grams paprika

2 tablespoons/30 milliliters dry red wine

10 feet/3 meters hog casings or 20 feet/6 meters sheep
   casings, soaked in tepid water for at least 30 minutes
   and rinsed

1. Combine the meat with the salt and DQ Curing Salt #2 or Insta Cure #2 and grind through the small die into the bowl of standing mixer set in ice (see Note below).

2. Dissolve the Bactoferm in the distilled water and add it, along with the rest of the ingredients, to the meat. Using the paddle attachment, mix on the lowest speed to incorporate all the ingredients, 1 to 2 minutes.

3. Stuff the sausage into the casings, and twist into 10-inch/25-centimeter links. Using a sterile pin or needle, poke holes all over the casings to remove any air pockets and facilitate drying.

4. Hang the sausage at room temperature, ideally 85 degrees F./29 degrees C. for 12 hours to "incubate" the bacteria; the beneficial bacteria will grow and produce more lactic acid in warmer temperatures.

5. Hang the sausage to dry (ideally at 60 degrees F./18 degrees C. with 60 to 70 percent humidity) until completely firm and/or it has lost 30 percent of its weight, 6 to 8 days if using sheep casing, 12 to 18 days if using hog casing.

6. *Optional:* For cooked peperone, hot-smoke (see page 74) it lightly at 180 degrees F./82 degrees C. to an internal temperature of 145 degrees F./62 degrees C., about 2 hours.

Yield: Twelve 10-inch/25-centimeter sticks if using hog casings,
twenty-four 10-inch/25-centimeter sticks if using sheep casings

[NOTE: See pages 105–115 for a detailed description of the basic grinding, mixing, stuffing, and cooking techniques.]

SOPPRESSATA

Soppressata, similar to salami, is another dry-cured, fermented sausage, though typically the meat is not as finely ground, so it has a less uniform, more speckled appearance, with some bigger chunks of fat in the slice. Because the fat in this sausage should be distinct from the meat, it's especially important during the grinding and mixing stages to keep the fat as cold as possible, to avoid smearing the fat on the meat.

A specialty in southern Italy, notably in Calabria and Basilicata, soppressata is sometimes smoked and often pressed, thus its name. Seasonings vary, and there are infinite variations and hundreds of published recipes in Italian cookbooks. This is a straightforward version—seasoned with garlic and pepper, red pepper flakes, and white wine—and not too hot, but feel free to double the amount of red pepper if you like a spicy sausage. Soppressata is traditionally made with larger hog middles (about 3 inches/7.5 centimeters in diameter), but hog casings will work as well.

1 pound/450 grams pork back fat, diced (see Note 1 below)

4 pounds/1800 grams boneless pork shoulder, diced

1 tablespoon/10 grams Bactoferm F-RM-52 (live starter culture; see Sources, page 303)

¼ cup/60 milliliters distilled water

¼ cup/56 grams kosher salt

1 teaspoon/7 grams DQ Curing Salt #2 or Insta Cure #2 (see page 105)

3 tablespoons/30 grams dextrose

1 teaspoon/3 grams ground white pepper

1 teaspoon/6 grams minced garlic

1 teaspoon/2 grams hot red pepper flakes

¼ cup/60 milliliters Pinot Bianco or comparable dry white wine

**12 to 14 inches/30 to 35 centimeters hog middle or 10 feet/
3 meters hog casings, soaked in tepid water for at
least 30 minutes and rinsed**

1. While the fat is very cold, grind through the medium die into a bowl set in ice (see Note 2 below). Chill while you grind the meat through the large die. Combine the ground meat and fat in the bowl of a standing mixer and refrigerate while you ready the culture and the remaining ingredients.

2. Dissolve the Bactoferm in the distilled water and add it, along with the remaining ingredients, to the meat. Using the paddle attachment, mix on the lowest speed until the seasonings are thoroughly distributed, 1 to 2 minutes.

3. Stuff the sausage into casings. Tie the ends of the hog middle, if using. Or, if using hog casings, twist into 8-inch/20-centimeter links. Using a sterile pin or needle, prick the casings all over to remove any air pockets and facilitate drying.

4. Hang the sausage at room temperature, ideally 85 degrees F./29 degrees C., for 12 hours to "incubate" the bacteria; the beneficial bacteria will grow and produce more lactic acid at a warmer temperature.

5. Hang the sausage (ideally at 60 degrees F./15 degrees C. with 60 to 70 percent humidity) until completely dry or until it's lost 30 percent of its weight. The time will differ depending on the size of the casings you use and your drying conditions–roughly 2 to 3 weeks.

Yield: About 3 pounds/1.5 kilograms sausage; one 14-inch/35-centimeter sausage
if using a hog middle, eight 8-inch/20-centimeter links if using hog casings

[NOTES: 1. If you choose to freeze your pork and fat, do so 2 to 3 weeks before making this sausage, according to the instructions on pages 177–178. Thaw the meat in the refrigerator for 1 to 2 days. 2. See pages 105–115 for a detailed description of the basic grinding, mixing, stuffing, and cooking techniques.]

COPPA

When we first published this recipe, coppa was scarcely known in America and only a handful of places sold it. And because so few chefs broke down whole hogs at their restaurants, few had access to the actual coppa muscle that runs from the neck over the shoulder and segues into the loin. American hog breakdown bifurcates this muscle. But so much has changed in America, and so quickly! Now coppa, also referred to as capocollo, is on virtually every charcuterie and salumi platter in America.

Brian created this version of coppa to imitate the whole muscle version since so few had access to a whole coppa. And we retain it here as a vestige of America's past and as an excellent way for home dry-curers to mimic actual coppa. This recipe uses chunks of shoulder (from same general area of the pig and with a similar fat content), seasoned and packed tightly into large beef middles. Rubbing the seasoning vigorously into the meat to make it sticky and packing it tightly to avoid any air pockets is critical.

In Tuscany, common seasonings include garlic, orange and lemon zest, cinnamon, and caraway. Here we include recipes for both spicy and sweet coppa. As with all these larger dried sausages, it should be sliced paper-thin and served before the meal, as an antipasto.

For more on the coppa cut and dry curing it in the traditional Italian manner, see our book *Salumi*.

5 pounds/2.25 kilograms boneless pork shoulder butt, well trimmed of fat and sinew, cut into chunks about 3 inches/7.5 centimeters square (see Note 1 below)

THE DRY CURE
about ½ cup/125 grams kosher salt
2 ½ tablespoons/25 grams dextrose
1 teaspoon/7 grams DQ Curing Salt #2 or Insta Cure #2 (see page 105)

FOR SPICY SAUSAGE
Hot paprika, preferably Hungarian to coat the meat (about 4 tablespoons/32 grams)
1 tablespoon/9 grams cayenne pepper

FOR SWEET SAUSAGE
3 tablespoons/40 grams sugar
2 tablespoons/20 grams freshly ground black pepper
1 tablespoon/8 grams ground coriander
2 teaspoons/12 grams minced garlic
1 teaspoon/4 grams ground mace

1 teaspoon/4 grams ground allspice

¾ teaspoon/3 grams ground juniper berries

Beef middles, about 18 to 20 inches/45 to 50 centimeters long,
soaked in tepid water for at least 30 minutes and rinsed

1. Combine the salt, dextrose, and DQ Curing Salt #2 or Insta Cure #2. Rub the meat all over with half the cure mixture and place in a single layer in a nonreactive baking pan or sheet. Cover with plastic wrap and refrigerate for 3 days.

2. Rub the meat with the remaining cure mixture and pack tightly in a single layer in a nonreactive pan. Cover with plastic wrap and refrigerate for another 3 days.

3. Remove the meat from the refrigerator, rinse under cold water, and pat dry with paper towels. Combine all the spicy or sweet seasonings and rub the pork well with the seasoning until the surface becomes sticky, about 5 to 8 minutes.

4. Pack the meat tightly by hand into the casings (see Note 2 below). Prick any air pockets with a sterile pin or needle before tying the ends.

5. Hang the coppa at room temperature for 12 hours.

6. Dry the coppa (ideally at 60 degrees F./15 degrees C. with 60 to 70 percent humidity) until completely firm throughout, 3 to 4 weeks.

Yield: One 18- to 20-inch/45- to 50-centimeter coppa

[NOTES: 1. If you choose to freeze your pork, do so 2 to 3 weeks before making this sausage, according to the instructions on pages 177–178. Thaw the meat in the refrigerator for 2 to 3 days. 2. See pages 105–115 for a detailed description of the basic grinding, mixing, stuffing, and cooking techniques.]

SPANISH CHORIZO

One of the best dry-cured sausages there is, for its rich flavor and versatility. It's excellent sliced and eaten as a canapé. It can be sautéed and added to beans or stews or eggs. It's a key ingredient in paella. Chorizo is often smoked but it doesn't have to be. Its distinctive characteristic is the paprika, so be sure to use smoked Spanish paprika, as we do here, or fresh Spanish or Hungarian paprika, paprika that hasn't been hanging out in the spice rack for several years.

5 pounds/2.25 kilograms boneless pork shoulder butt, diced
(see Note 1 below)

1/4 cup/56 grams kosher salt

1 teaspoon/7 grams DQ Curing Salt #2 or Insta Cure #2
(see page 105)

1 tablespoon/10 grams dextrose

1 tablespoon/10 grams Bactoferm F-RM-52 (live starter culture;
see Sources, page 303)

1/4 cup/60 milliliters distilled water

2 tablespoons/16 grams smoked hot Spanish paprika (*pimentón*)

2 tablespoons/16 grams ancho chile powder

1 1/2 teaspoons/5 grams cayenne pepper

2 tablespoons/36 grams minced garlic

10 feet/3 meters hog casings, soaked in tepid water for at
least 30 minutes and rinsed

1. Combine the pork with the salt, DQ Curing Salt #2 or Insta Cure #2, and dextrose. Grind through the large die into the bowl of standing mixer set in ice (see Note 2 below).

2. Dissolve the Bactoferm in the distilled water and add it, along with the remaining ingredients, to the pork. With the paddle attachment, mix on low speed about 1 minute to incorporate all the ingredients.

3. Stuff the sausage into the hog casings and use string to tie into 12-inch/30-centimeter loops. Using a sterile pin or needle, prick the casings all over to remove any air pockets and facilitate drying.

4. Hang the sausage (ideally at 60 degrees F./15 degrees C. with 60 to 70 percent humidity) until it feels completely stiff throughout and/or it's lost 30 percent of its weight, 18 to 20 days.

Yield: About 3 pounds/1.5 kilograms sausage; 5 large rings

[NOTES: 1. If you choose to freeze your pork, do so 2 to 3 weeks before making this sausage, according to the instructions on pages 177–178. Thaw the meat in the refrigerator for 1 to 2 days. 2. See pages 105–115 for a detailed description of the basic grinding, mixing, stuffing, and cooking techniques.]

HUNGARIAN SALAMI

Brian was inspired to make a Hungarian sausage because of his Hungarian mother-in-law. This Hungarian-style salami is like traditional salami in texture but with the additional flavors of garlic and paprika. Cold-smoking the sausage (as Spanish chorizo often is) adds a depth of flavor as well as a mold-inhibiting acidic coating to the casing. If you don't have the capacity to cold-smoke, though, this can be dried in the manner of the Tuscan Salami (see page 181).

Paprika plays a big role in this sausage. We recommend that you stay away from generic brands; use good Hungarian paprika or any high-quality paprika such as La Vera. And be sure the paprika you use is very fresh; if you don't remember when you bought it, take that as an indicator of how fresh it is.

1.5 pounds/675 grams pork back fat, diced (see Note 1 below)

2 pounds/900 grams boneless lean pork shoulder butt, diced

1.5 pounds/675 grams boneless lean beef chuck, diced

¼ cup/56 grams kosher salt

1 teaspoon/7 grams DQ Curing Salt #2 or Insta Cure #2
(see page 105)

1 tablespoon/10 grams Bactoferm F-RM-52 (live starter culture;
see Sources, page 303)

¼ cup/60 milliliters distilled water

6 tablespoons/48 grams Hungarian paprika

2 tablespoons/20 grams dextrose

2 tablespoons/20 grams black peppercorns, soaked overnight
in water to cover generously

1 teaspoon/8 grams minced garlic

2 tablespoons/30 milliliters dry white wine

12 feet/3.6 meters hog casings or 3-foot/1-meter hog middles,
soaked in tepid water for at least 30 minutes and rinsed

1. Grind the fat through the large die into a mixer bowl set in ice (see Note 2 below). Refrigerate.

2. Combine the pork, beef, salt, and DQ Curing Salt #2 or Insta Cure #2, and grind through the small die into the mixer bowl set in ice. Refrigerate.

3. Dissolve the Bactoferm in the distilled water. Combine the meat, fat, Bactoferm, dry milk, paprika, dextrose, peppercorns, and garlic in the bowl of a standing mixer. Using the paddle attachment, mix on the lowest speed until the ingredients are evenly combined, about 1 minute. Add the wine and mix for 1 more minute.

4. Pack the mixture tightly into a nonreactive pan. Cover with plastic wrap, pressing it down against the meat so there are no air pockets between the meat and plastic. Refrigerate for 2 days.

5. Stuff the sausage into the casings. Twist into 12-inch/30-centimeter links if using hog casings. Or tie the ends of the hog middles. Using a sterile pin or needle, prick the casings all over to remove any air pockets and facilitate drying. Hang at room temperature for 12 hours.

6. Cold-smoke the sausage (see page 74) at 70 degrees F./21 degrees C. for 4 to 6 hours.

7. Hang the sausage (ideally at 60 degrees F./15 degrees C. with 60 to 70 percent humidity) for 18 to 20 days until it feels completely firm and/or it has lost 30 percent of its weight.

Yield: 4 pounds/1.75 kilograms sausage; 12 to 14 10-inch/25-centimeter
links if using hog casings, 2 long sausages if using hog middles

[NOTES: 1. If you choose to freeze your pork and fat, do so 2 to 3 weeks before making this sausage, according to the instructions on pages 177–178. Thaw the meat in the refrigerator for 1 day. 2. See pages 105–115 for a detailed description of the basic grinding, mixing, stuffing, and cooking techniques.]

SAUCISSON SEC

This is one of the easiest sausages to make, a very pure kind of dry sausage and one of our favorites for that reason—dry-cured pork, with just a little bit of garlic and pepper (and no added bacteria culture). It tastes like the French countryside; eat it with a good Burgundy and a great baguette. Because of the very subtle seasonings, the quality of the pork is more noticeable here than in heavily spiced sausages.

4½ pounds/2 kilograms boneless pork shoulder butt, diced
(see Note 1 below)
8 ounces/225 grams pork back fat, diced
¼ cup/56 grams kosher salt
1 tablespoon/10 grams coarsely ground black pepper

1½ tablespoons/15 grams sugar

1 teaspoon/7 grams DQ Curing Salt #2 or Insta Cure #2

(see page 105)

1 tablespoon/18 grams minced garlic

12 feet/3.6 meters hog casings or 3-foot/1-meter hog middles,

soaked in tepid water for at least 30 minutes and rinsed

1. Grind the pork and fat through the large die into a bowl set in ice (see Note 2 below).

2. Combine the meat with the remaining ingredients in the bowl of a standing mixer. Using the paddle attachment, mix on the lowest speed until the ingredients are evenly combined, about 1 minute.

3. Stuff the mixture into the casings and twist into 12-inch/30-centimeter links if using hog casings. Or tie the ends of the hog middles. Prick the casings all over with a sterile pin or needle to remove any air pockets and facilitate drying.

4. Hang the sausage (ideally at 60 degrees F./15 degrees C. with 60 to 70 percent humidity) until it feels completely stiff throughout and/or it has lost 30 percent of its weight, 18 to 20 days for links, a month or more for large sausages.

Yield: About 3 pounds/1.5 kilograms sausage; about ten 12-inch/28-centimeter links if using hog casings, or 1 long sausage if using hog middles

[NOTES: 1. If you choose to freeze your pork and fat do so 1 to 2 weeks before making this sausage, according to the instructions on pages 177–178. Thaw the meat in the refrigerator for 1 day. 2. See pages 105–115 for a detailed description of the basic grinding, mixing, stuffing, and cooking techniques.]

LANDJAGER

A simple cold-smoked, completely dried sausage. Its name translates as "land soldier," as it was an excellent portable food for Europeans during times of war, and is not unlike American beef jerky in its portability. Traditionally, the sausages are pressed between wooden boards that have one-inch channels carved in them, giving the sausages a squared-off rectangular shape. They are a very dark brown and so dry that they will snap in half, so they're best thinly sliced to serve.

2 pounds/900 grams boneless pork shoulder butt, diced (see Note below)

3 pounds/1350 grams boneless beef round, diced

¼ cup/56 grams kosher salt

1 teaspoon/7 grams DQ Curing Salt #2 or Insta Cure #2 (see page 105)

1½ tablespoons/15 grams dextrose

1¼ ounces/35 grams Fermento, see Sources, page 303)

¼ cup/60 milliliters distilled water

¾ teaspoons/3 grams garlic powder

2 teaspoons/6 grams caraway seeds

4 teaspoons/12 grams freshly ground black pepper

1 teaspoon/3 grams coriander seeds, toasted and ground

10 feet/3 meters hog casings, soaked in tepid water for at least 30 minutes and rinsed

1. Combine the pork and beef with the salt, DQ Curing Salt #2 or Insta Cure #2, and dextrose. Grind the meat mixture through the small die into the bowl of an electric mixer set in ice.

2. Dissolve the Fermento in the distilled water. Add to the meat, along with the remaining ingredients. Using the paddle attachment, mix at the lowest speed until the ingredients are incorporated and the meat has a tacky appearance, 1 to 2 minutes.

3. Stuff the sausage into the hog casings, and twist into 5-inch/12-centimeter links. Prick any air pockets with a sterile pin.

4. Line a baking sheet with parchment paper and lay the sausages out in a single layer. Cover with parchment paper, place another baking sheet on top, and weight that baking sheet with about 5 pounds/2 kilograms. Refrigerate the sausage for 2 days.

5. Cold-smoke the sausages (see page 74) for 4 hours.

6. Hang the sausages to dry (ideally about 60 degrees F./15 degrees C. with a humidity of 60 percent or higher) for 12 to 15 days, until hard.

Yield: About twenty 5-inch/12-centimeter links

[NOTE: If you choose to freeze your pork, do so 2 to 3 weeks before making this sausage, according to instructions on pages 177–178. Thaw the meat in the refrigerator for 1 to 2 days.]

Hams

The dry-cured ham elicits a reverence perhaps unmatched by any other single charcuterie item. By dry-cured ham, we mean the whole back leg of a mature hog packed in salt for weeks, then rinsed and hung to dry for many months or even years. In France, it can be called *jambon cru*, but it may be called jambon de Bayonne if the ham comes from a hog grown in that Basque region, where it was fed on chestnuts and hickory nuts. In Italy, it is prosciutto di Parma if it comes from a hog raised in Emilia-Romagna and fed on cows' whey, a by-product of Parmigiano-Reggiano, the region's famous cheese. In all these regions, local artisans have ham making down to a virtual science, with master salters to cure the meat, and trained inspectors who slide long needles made of porous, odor-absorbing bone into the ham to check for quality. Most European countries have their own variations on the technique—some use a little smoke, some use seasonings, but in principle it's the same thing. Dry-cured ham sliced very thin and eaten straight off the knife, or with crusty toast drizzled with olive oil.

Writer Peter Kaminsky traveled the globe in search of the perfect ham for his book *Perfect Pig*, and he speaks about dry-cured ham with the reverence typically given fine wines, noting nuances of flavor and the fat quality. His favorite is the Iberian ham, *jamón ibérico*, from a dark breed native to Spain, descended from wild boars from southwestern environs. They're free-range pigs and fatten themselves on acorns from the abundant oaks there.

"Iberian is the queen," says Kaminsky. "The pigs eat grass and mast. They're raised on what looks to be a three-million-acre golf course. It's one of the greatest tastes in the world. It's silky, it's got some funkiness to it, salty, slightly bitter at the end. It's complex and balanced."

The best generally are those hams that can be aged more than a year, a capability often determined by the amount of fat—the more fat, the longer the ham can dry. The fat affects flavor and texture too. Kaminsky says the diet of acorns results in soft fat with a low melting point, a quality American Smithfield hams used to have as well (when they were fed on peanuts). A long drying time, like any aging, results in changes of flavor.

The general method for any dry-cured ham is standard. The hams are packed in salt and sometimes stacked on top of each other, their weight pressing water out, and typically cured for about a day for every 2 pounds/1 kilo of weight. The salt, as always, reduces bacterial growth that would cause spoilage, and the drying reduces the water content of the meat, further retarding bacteria and concentrating the flavor.

The quality of dry-cured ham may be most reliant on the quality and diet of the hog. You can dry-cure any fresh ham, it's not difficult, but if the hog is from an American commercial grower, the ham is not going to taste anywhere near as fine as it would were it from a hog raised on grass and acorns and allowed to grow big and fat (the Iberian hogs can grow to four hundred pounds, while American factory hogs average about two hundred and fifty pounds at slaughter). What makes hams so special is the region, the diet of the pig, and, many believe, the air itself in this or that region of Europe. Size also increases the required drying time, thus enhancing flavor.

Some hams, such as those from Westphalia in Germany, are smoked with local wood. (Kaminsky notes that smoke is used proportionately to latitude: the farther north—as the climate grows increasingly damp—the more smoke is used.) Some pack the exposed flesh in lard or dust it with pepper to keep it from drying too much and to keep bugs off it. Some producers encourage molds and believe they help the flavor; others can take it or leave it.

Done well, with a good hog and respect for the region, it can be one of the greatest flavors in the world.

SALTED AIR-DRIED HAM

This ham is in the style of the most famous hams, prosciutto di Parma and San Daniele, Bayonne, and Serrano. It's the most simple kind of dry-cured ham. Anyone can do the curing, but the quality of the end result is entirely dependent on the hog, where it lived, what it ate, how fat it grew. We highly recommend this recipe and technique for anyone who has access to carefully grown or organically raised hogs.

It's important to keep the entire ham well covered with salt during its cure.

> **4 pounds/2 kilograms kosher salt, or as needed to coat the ham**
> **One 12- to 15-pound/6- to 7-kilogram fresh ham, skin on, aitch-bone removed**
> **½ cup/500 grams lard**
> **Cracked black pepper**
> **Cheesecloth**

1. Rub the salt heavily all over the ham, especially on the exposed flesh and around the exposed femur bone.

2. Place skin side down in a nonreactive roasting pan or plastic tub, cover with plastic wrap, and place another pan on top: weight the ham with about 10 pounds/5 kilograms (cans or clean bricks). Refrigerate for 1 day for each pound/500 grams, checking every couple of days to make sure all areas are still covered in salt. Pour off any excess water and add more salt if necessary. Avoid touching the ham with your bare hands too much; you may want to use disposable rubber gloves for sanitation (a little bit of bacteria can be a big problem).

3. On the last day of curing, the ham should feel firm and dense to the touch. If it does not, resalt as necessary and cure for another 1 to 3 days.

4. Wipe the remaining salt off the ham, rinse under cool water, and pat dry with paper towels. Spread the lard over the exposed meat and pack the cracked pepper onto the lard (the lard helps to keep the exposed flesh from overdrying and the pepper helps to keep bugs away). Wrap the ham in four layers of cheesecloth and tie with butcher's string.

5. Hang the ham in a cool, dry place (ideally 60 degrees F./15 degrees C. with 60 to 70 percent humidity) with good ventilation for at least 4 to 5 months, or as long as a year. The ham should lose almost half of its original weight. You will know it is ready when there isn't much give when you squeeze it. (You can also take a metal skewer and insert it in the center remove and smell—it should have a cured aroma. This takes practice.)

6. When the ham has dried, wipe off all the lard and carefully remove the rind with a sharp boning knife. Slice paper-thin, parallel to the bone with a sharp slicing knife.

Yield: One 9- to 11-pound/4- to 5-kilogram cured ham

BLACKSTRAP MOLASSES COUNTRY HAM

This is an American Southern-style ham that's cured and heavily smoked. Blackstrap molasses, available at health food stores and some grocery stores, is a residue of cane sugar refining, very syrupy and rich; it gives the skin of this ham a very dark color, nearly black, and the meat a smoky-sweet flavor. This style of ham, one that's cured, then cold-smoked before drying, is common in the United States and in the northern climes of Europe. The best-known smoked and dry-cured hams today may be those of the German region of Westphalia (once a province of Prussia), a region that's been famed for its cold-smoked hams for two millennia.

This ham can be a little more tricky than its cousin, the American-style glazed ham (page 91). Whereas that ham is brined and completely cooked, this ham is dry-cured, then heavily smoked, then dried. Ideally, when fully dried the interior will be as dense as any dry-cured ham. If it is not dried through to the center, it is still delicious cooked, especially in soups, stews, braised greens–any dish that would benefit from the smoked ham flavor.

3 pounds/1.3 kilograms kosher salt

1 tablespoon/28 grams DQ Curing Salt #2 or Insta Cure #2
(see page 105)

1 pound/450 grams dark brown sugar (about 2¼ packed cups)

2¼ cups/560 milliliters blackstrap molasses

1 cup/250 milliliters dark rum

1 tablespoon/10 grams grated or minced fresh ginger

1 teaspoon/3 grams cayenne pepper

½ teaspoon/1 gram coriander seeds, toasted and ground
(see Note page 51)

1 tablespoon/8 grams juniper berries, crushed with the side
of a knife

One 12- to 15-pound/6- to 7-kilogram fresh ham, skin on,
aitch-bone removed

1. Combine the salt and DQ Curing Salt #2 or Insta Cure #2 in a medium bowl and stir. Add all the remaining ingredients except the ham and stir to combine.

2. Rub the cure mixture all over the ham, giving extra attention to the area around the exposed bone. Place in a nonreactive container big enough to contain the ham and the liquid that will be drawn out by the salt cure. Using a platter or a board and some type of weights, weight the ham with about 15 pounds/7 kilograms. Refrigerate for 12 to 15 days–1 day per 1 pound/450 grams); turn the ham and redistribute the cure as necessary after 6 days.

3. Remove the ham from the cure and rinse it under cold water. Soak it in cold water for 8 hours to remove residual salt from the surface.

4. Set the ham on a rack and refrigerate it, uncovered, overnight.

5. Cold-smoke the ham (see page 74), for 18 hours at 60 degrees F./15 degrees C.

6. Hang the ham in a cool, dark place (ideally 60 degrees F./15 degrees C. with 65 to 70 percent humidity) to dry for 7 weeks.

Yield: One 10-pound/5-kilogram smoked ham

BRESAOLA

This air-dried beef is common in the mountainous regions of northern Italy. A very lean, intensely flavored preparation, it's usually sliced paper-thin and drizzled with olive oil and lemon juice. You might also serve it with shaved Parmigiano-Reggiano, some greens, and thin slices of toasted baguette. If possible, use grass-fed or organically raised beef.

THE SPICE CURE
¼ cup/56 grams kosher salt
2 tablespoons/30 grams sugar
1 ½ teaspoons/5 grams freshly ground black pepper
1 tablespoon/6 grams chopped fresh rosemary
2 teaspoons/6 grams fresh thyme leaves
5 juniper berries, crushed with the side of a knife

One 3-pound/1.5-kilogram beef eye of the round roast, no
more than 3 inches/7.5 centimeters in diameter,
trimmed of all visible fat, sinew, and silverskin

1. Combine all the spice cure ingredients in a spice or coffee grinder and grind to a fine powder.

2. Rub half the spice cure all over the meat, rubbing it in well. Place in a 2-gallon/8-liter Ziploc bag or a nonreactive container and refrigerate for 7 days, turning it every couple of days.

3. Remove the beef from the liquid (discard it) and rub in the remaining spice cure. Return to the refrigerator for 7 more days.

4. Rinse the beef thoroughly under cold water to remove any remaining spices and pat dry with paper towels. Set on a rack on a baking sheet uncovered at room temperature for 2 to 3 hours.

5. Tie the beef with butcher's twine. Hang the meat (ideally at 60 degrees F./15 degrees C. with 60 to 70 percent humidity) for about 3 weeks. The meat should feel firm on the outside and silky smooth when sliced.

Yield: 2 pounds/1 kilogram bresaola; about 30 appetizer servings

LARDO AND CURED PORK BELLY

Lardo is cured pork back fat, pure unadulterated fat. If you use belly instead, you'll have striations of meat. Both are excellent–provided you use excellent pork, preferably from a small farmer who raises his or her hogs naturally. Do not use any other kind of pork; it's simply not worth it from a flavor standpoint, and you're likely to wonder what all the fuss is about. Either of these preparations couldn't be easier. The fat or pork belly is simply cured with a dry rub of salt, sugar, and herbs, then hung to dry. Sliced thin, it can be eaten plain, with some olive or truffle oil and crusty bread or toast. It's delicious cooked as well.

Because light damages fat, and because these pieces are mainly fat, it's important to cure them in a way that prevents as much exposure to light as possible. Then store the cured lardo or belly wrapped in plastic wrap followed by foil.

> ½ recipe Basic Dry Cure (page 39)
> 3½ pounds/1.5 kilograms pork back fat or pork belly in
>   one piece (about 12 inches/30 centimeters long by
>   6 inches/15 centimeters wide and 1¼ inches/
>   3 centimeters thick), skin removed
> 6 bay leaves
> 2 bunches fresh thyme
> ¼ cup/40 grams black peppercorns
> Cheesecloth (optional; see step 3 below)

1. Sprinkle a quarter of the dry cure mix into a nonreactive baking pan large enough to hold the pork. Place the pork on top, and sprinkle the remaining cure mix over it. Distribute the remaining ingredients on top of the pork and cover with plastic wrap, then wrap in foil to protect the pork from light.

2. Weight the pork with about 10 pounds/5 kilograms and refrigerate for 10 to 12 days, turning it and rubbing it twice to redistribute the cure. The pork is cured when it feels uniformly dense and stiff.

3. Remove the pork from the pan, rinse thoroughly under cold water, and pat dry. If you don't have an enclosed drying box or room, wrap the pork in a few layers of cheesecloth. Poke a hole in one corner of the pork and tie a string through it to make a loop for hanging.

4. Hang the pork to dry in a cool, dark humid place (ideally about 60 degrees F./15 degrees C. with 60 to 70 percent humidity) for 18 to 24 days.

Yield: About 30 appetizer servings

6.

THE CINDERELLA MEAT LOAF

. . . .

Pâtés and terrines, perhaps because they are served in elegant restaurants and sold at gourmet shops, and often look so cool, are thought to be unmercifully difficult to make. Some are (for example, *pâtés en croûte*). But the basic country pâté, which can be every bit as satisfying as a fancy-pants *galantine de canard*, is scarcely more difficult than your average made-from-scratch meat loaf. The following recipes run the gamut from basic to elaborate, but the bottom line is that the pâté represents a basic technique that the home cook should enjoy.

Pâtés and terrines, broadly speaking, are essentially big sausages cooked in some sort of mold, either earthenware or porcelain (*en terrine*), in dough (*en croûte*), or in skin (galantines and ballotines). Without a mold, they're meat loaf. (In this chapter, we use the words *pâté* and *terrine* interchangeably. Technically, though, *terrine* is short for *pâté en terrine*.)

Pâtés made of meat and fish date to antiquity. They remained popular from the Middle Ages up through the 1900s, and today they continue to be a part of restaurant cooking but are largely neglected in the home kitchen.

We tend to associate pâtés with French cooking because French haute cuisine showed how special they might be, given some care and finesse (compare the elegant *pâté en croûte* with the workman-like English pork pie, for instance). In America, though, as with so much French-inspired haute cuisine, pâtés became somewhat debased and misinterpreted by cooks who didn't understand what the greatness of the pâté was all about–a smooth suspension of fat in meat, beguiling spices, forceful seasoning, and a dynamic appearance. Americans in general have the impression that a pâté is a meat spread or a liver spread, or that it must include foie gras. While foie gras can make a pâté of the highest order, in the hands of the right cook, chicken and salmon make equally exquisite pâtés. In France, pâtés, en terrine and en croûte, are available at markets everywhere, very much a lively part of culinary commerce. There are indications that it's returning in popularity here as more chefs offer their own pâtés at their restaurants. And now the pâté, like so many great dishes nearly lost, may be ready for a resurgence in the home kitchen.

The only special equipment you need to make a pâté en terrine is a meat grinder or food processor, depending on the type of pâté. As a pâté can be shaped into a log, wrapped in plastic or cheesecloth, and poached, you don't even need a terrine mold, the porcelain or earthenware vessel in which a pâté is traditionally cooked. Moreover, because a pâté must be made well in advance of serving, and serves a lot of people, it's a great dish for entertaining, either set out on a cutting board with some mustard or sliced and served as an opening course. Leftovers go beautifully with salad and some good crusty bread.

And a pâté can be very inexpensive to put together, especially if you use trimming and/or inexpensive meats (one hallmark of a talented chef). If you already have the spices in hand, a pâté de campagne serving ten to fifteen people can be made for less than $10.

Basic Pâté Technique

While a pâté is not always an emulsion, often the fat is distributed evenly throughout the meat so that it's present without calling attention to itself. Even in pâtés in which the fat is coarsely ground and distinct from the meat, care must be taken to maintain the suspension by keeping your bowls, grinder or processor bowl, blades, and ingredients cold. Otherwise, the fat may liquefy and melt out of the pâté. This is especially important in pâtés that are pureed in a food processor. You may not believe this until you pull a terrine from the oven and see it shrunken and floating in a pool of clear fat, the fat and meat having separated. It is edible, but it will be rubbery rather than smooth, dry instead of rich.

The term *forcemeat*—from the French *farcir*, to stuff—means ground or pureed meat, fish, or vegetables. Forcemeats for pâtés are traditionally divided into four categories, each of which denotes a specific technique: country, gratin, straight, and mousseline.

For a country pâté, the meat is not pureed, but ground for a coarse texture. The word *gratin* here indicates that some or all of the meat has been seared over high heat and cooled before being combined with the rest of the ingredients, adding the roasted flavor to a pâté that will otherwise only be gently cooked. For a straight forcemeat, uncooked pork, pork fat, and often some other meat—duck or venison or rabbit—are ground, mixed, and pureed in a food processor. And a mousseline forcemeat is one made by pureeing ground chicken, veal, or fish with egg white and cream. It's very stable and the easiest to make at home.

Escoffier believed that any type of meat could be well served by the mousseline method, which results in a forcemeat that, in terms of delicacy, he said, "cannot be surpassed." (And that was back in the day when cooks pounded the meat by hand and laboriously pushed it through a sieve before stirring in the egg white and cream.)

The main stages in preparing an excellent pâté (as with sausage making), are seasoning, grinding, mixing, testing for seasoning, and cooking. The meat, fat, and seasonings are ground, then either blended in an electric mixer or pureed in a food processor. Some "interior garnish," such as nuts, mushrooms, or herbs, may be folded in for texture, flavor and visual impact, then the mixture is baked in a water bath or poached, and chilled.

SEASONING

The chunks of meat and fat are always seasoned well in advance, to allow the seasoning to do its work, and refrigerated. The first and last stages of most cooking, and all the way in between, are seasoning. Seasoning is the critical factor in a finished pâté. Because pâtés are served cold, the seasoning must be aggressive.

THE IMPORTANCE OF TEMPERATURE

Once you are ready to grind your meat and fat, temperature becomes a critical factor in the success of a terrine, and it will remain a factor until the terrine goes into the oven. In order to combine the meat and fat perfectly, they must be very cold. If they become warm, the fat can soften or melt, and ultimately you can wind up with a broken forcemeat, just as you can wind up with a broken hollandaise, if you don't pay attention to temperature.

GRINDING

Except for that in a mousseline forcemeat, all meat for pâtés is passed through a meat grinder. Usually the finest die is appropriate, but some cooks prefer to grind first through a large die, then a smaller one (a process technically called progressive grinding). If you are looking for a more rustic terrine, as with the pâté de campagne, you might choose to grind some of your meat only through the large die.

Until your pâté goes into the oven, you must do all that you can to keep the meat cold. Don't let your ingredients and tools get warm. You don't have to be fanatical about it–moving your KitchenAid out on the back deck in February snow–but do be slightly paranoid about it. Chill your grinder attachment in the freezer for an hour or more before grinding. When not working with any of the ingredients, keep them refrigerated. Set the bowl that will catch the ground meat in a larger bowl of ice. It's handy to use the mixer bowl to catch the meat because often that's where it needs to be for the next step; also, metal conducts the cold well, as opposed to plastic, which doesn't.

If you need to stop after grinding, for whatever reason, return the meat to the refrigerator or freezer until you're ready to proceed. If you use the freezer, though, don't let the meat freeze solid–crunchy is acceptable, but not frozen solid.

MIXING

For forcemeats that aren't pureed in a food processor (as those for most pâtés are), such as most sausage forcemeats, mixing the ground meat for a minute or two using the paddle attachment of your mixer, or stirring vigorously with a wooden spoon, helps to develop the myosin protein in the muscle. You will see the ground meat change from a hamburger-like mound to a sticky-looking paste; the stickiness is a result of the protein development. Developing the myosin protein helps the forcemeat hold together. But you don't need or want to mix the meat too long, because that mechanical action heats the meat and fat. If you're using a wooden spoon, it helps to put the mixing bowl in ice to keep the mixture cold. This mixing also helps to distribute seasoning more evenly as well as to incorporate any liquid.

PUREEING

Many pâté mixtures are pureed in the food processor, which results in a fine smooth texture in the finished pâté. Freeze your processor bowl and blade before pureeing the ground meat and fat. Puree only as long as necessary, usually until the mixture is smooth and uniform. Again, temperature is the key concern here, given that the whipping blade and warm motor heat up the meat.

PANADE

A panade is a mixture of starch, often simply plain white bread, and liquid that helps to bind a pâté or terrine mixture. It may also contain eggs, and the liquid is often cream, both of which further serve to bind and enrich.

SEASONING AND THE QUENELLE TEST

Classic seasonings are salt, of course, and a mixture of sweet spices, such as nutmeg and clove, or quatre épices (see page 143). Check your seasoning before you commit your pâté to the oven by cooking a spoonful of the mixture. It's best to wrap it in plastic and gently poach it in water (160 to 170 degrees F./71 to 76 degrees C.), because that will best reflect the final dish; if you fry it over direct heat like a mini-burger, it will cook at too high a temperature, losing its fat and developing a fried flavor that the pâté won't have, making it difficult to evaluate the seasoning. Taste the poached sample. Does it need a little more salt? If it doesn't seem to, think again—the pâté will be served cold, and cold food needs more aggressive seasoning than hot. So your hot pâté test should taste slightly oversalted. If it needs any other flavor adjustments, now is the time to act.

INTERIOR GARNISH

Often pâtés are enhanced by other whole or chopped ingredients mixed into the pureed force-meat called interior garnishes. A large garnish such as a seared duck breast may be laid in the center of the pâté when you fill the mold, or smaller items, such as mushrooms, diced ham, tongue, duck confit, pistachios, or herbs, can be folded into the forcemeat. These garnishes add flavor, as well as visual and textural contrast. It's harder to make cold food taste wonderful than it is hot food, so the more visually ravishing a pâté–say with elegant red cherries, green pistachios, black trumpet mushrooms surrounding the oval of a duck breast–the more likely your guests will be tempted and pleased.

THE VESSEL ITSELF

It almost doesn't matter. There are countless rectangular molds made of various materials in all sizes. The first pâtés were baked in pastry, until the pastry guild said the charcuterie guild was cutting in on its territory and the charcutiers of the Renaissance were forced to put their loaves of meat, their pâtés, into earthenware vessels–terrines. You can spend hundreds of dollars on fancy terrines with screw-down tops for pressing. Le Creuset's enamel-coated iron ones are the industry standard, but there are a wide variety at most kitchen stores.

Of course size and shape affect the outcome–you don't want a container so big that the outside of the pâté is overcooked while the center is raw. You wouldn't fill a stockpot with a nice forcemeat and expect it to cook well. Some terrine molds have an elegant shape, others less so. If you're serving a pâté out of the mold, you'll want it to look nice. But generally, the material in which you bake a pâté has little effect. As Brian puts it, "I bake in everything. Everything. And the end result is the same."

COOKING

The recipes here generally fill a 1½-quart/1.5-liter terrine mold. If your mold is smaller, you can either reduce the recipe accordingly (if your mold holds only a quart, reduce the quantities by two-thirds), or you can fill your mold, then wrap and poach the extra in water as you would a galantine (see step 11 on page 225).

Gentle heat is required for any terrine. If the forcemeat becomes too hot, it will break–the fat will separate out. You can cook a pâté without a mold simply by wrapping it securely in plastic and gently poaching it in water (160 to 170 degrees F./71 to 76 degrees C.) to the right temperature; usually 150 degrees F./65 degrees C., but if it includes poultry, 160 degrees F./71 degrees C.

Two types of terrine molds, a cast-iron enameled one with a lid and a porcelain mold. The shape and size of the mold will determine the cooking time and shape of the pâté, but as far as materials go, it really doesn't matter what type of mold you use. These two molds are ideal for pâtés made from meat or fish. The lid on the enameled one is convenient, and both of the molds are elegant items to have in your kitchen.

If you're using a terrine mold and intend to remove the pâté to serve it, line the mold with plastic wrap, leaving a couple of inches of overhang on the two long sides to fold over the top. (Dampen the inside of the mold, and the plastic wrap will stick in those corners and edges.) Thin strips of fat or bacon are sometimes used to enclose a pâté. Some cooks still favor this method, but there's enough fat in a pâté already, and generally solid fat eaten cold is not a delight. Only one of the recipes here call for this type of lining, but if you like the idea, any meat terrine can be lined using this traditional method.

With the mold thus lined, fill it with your forcemeat. It helps if you thwack it in, snapping it off the spoon or rubber spatula, to prevent air pockets. Pack the mixture down thoroughly, pushing it down with the edge of the spoon or spatula, then pressing it down and smoothing it out with the back of the spoon or the spatula.

Fold the plastic wrap over the top so the terrine is completely enclosed, then cover with the lid or with foil.

Place the terrine in a high-sided roasting pan and fill the pan with enough hot water (very hot tap water, 150 to 160 degrees F./65 to 71 degrees C. is best) to come halfway up the sides of the mold. Put the roasting pan in an oven preheated to 300 degrees F./150 degrees C.

Terrines take anywhere from 45 to 90 minutes, depending on their size. Start checking the temperature with an instant-read thermometer a few minutes or so before the time given in the recipe. When the interior of the terrine reaches the desired temperature (as a rule, a pork should reach 150 degrees F./65 degrees C., poultry 160 degrees F./71 degrees C.), remove it from the oven and set it on a rack (it will continue to cook when removed from the oven, the interior temperature rising by between five and ten degrees F., sometimes more, an effect called carryover cooking). Let cool to room temperature.

PÂTÉ EN TERRINE WITH PORK TENDERLOIN INLAY

Mise en place for a pâté en terrine: ground and pureed meat and fat in a bowl set in ice, pork tenderloin inlay, and a terrine mold lined with plastic wrap, with enough overhang to fold over the top. The chunks of ham, chopped pistachios, and chopped black truffles will be folded in as additional garnish.

The ham, pistachios, and truffles (the secondary, or random, garnish) are folded into the pureed meat and fat mixture while the mixture is kept cold.

The lined terrine mold is filled about halfway with the ground meat mixture, which is pressed down with a spatula to remove any air pockets.

The pork tenderloin, the primary garnish, is laid in the center of the mold, then the mold is filled with the remaining ground meat mixture, completely covering the tenderloin. The tenderloin should be perfectly centered in the pâté.

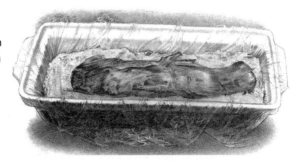

The plastic wrap overhang is folded over to enclose the pâté.

Pâtés must be cooked gently in a water bath, or they'll "break"—the fat will separate from the meat and the pâté may be dry and rubbery. The water bath prevents the temperature of the mold from going much above the boiling point. The filled, covered terrine mold is placed in a large roasting pan and the roasting pan filled with enough very hot tap water to reach two-thirds to three-quarters of the way up the mold. Then the roasting pan is carefully placed in the hot oven.

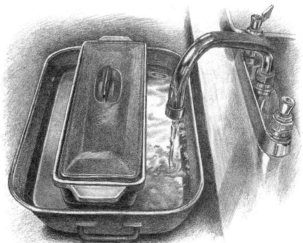

The cooked pâté should be weighted down while it cools to compact it and give it a smooth and uniform final texture, as well as to remove any remaining air pockets.

A cooked terrine should be pressed, weighted down, to ensure a perfectly even texture and density. Some terrine molds come with their own press, but most don't. To make your own, cut a piece of thick cardboard (or better, something stiffer, such as Plexiglas or wood) to the interior dimensions of the top of your mold. Wrap it in foil, place it on the terrine, and set a weight, about 2 pounds/1 kilogram on it (a brick wrapped in foil works well). When the terrine is cool enough to handle, refrigerate it overnight, brick on top.

How you serve a terrine depends on the type and the occasion, but when you're ready to serve it, whether as part of a winter holiday buffet or as a first course on a hot summer night, simply pop it out of the mold and carefully unwrap it. Place it on a cutting board. With a very sharp knife (preferably a slicing knife; don't use a serrated bread knife), cut it into uniform slices, about ⅜ inch/0.75 centimeter thick. It usually helps to run the knife under hot water, then dry it, before each slice, and a warm knife is mandatory for very delicate mixtures such as seafood mousselines or vegetable terrines.

Left on the cutting board, a sliced terrine can be served as an hors d'oeuvre. If you've made a beautiful terrine with an interior inlay of, say, a seared pork tenderloin and an interior garnish of green pistachios and shiitake mushrooms, you are probably going to feature it as a first course, with an appropriate sauce or condiment. In France, traditionally terrines were often served straight out of the mold (and still are), and that remains a good option for rustic dishes such as pâté grandmère (don't line the mold with plastic wrap, though!). Pâtés can also be the focal point of an elegant composed salad (a good vinaigrette is perfectly suited to most terrines). If you've made extra forcemeat, it can be wrapped in plastic, poached, and used in a composed salad.

Well wrapped in plastic, a terrine will keep for a week to ten days in the refrigerator.

Country Pâtés

Country pâtés (or country forcemeats) are distinguished by their coarse texture. The meat is ground but not pureed.

Straight Pâtés

Straight pâtés (or straight forcemeats) are distinguished by their smooth texture; the meat is first finely ground, then pureed in a food processor.

Gratin Pâtés

When used in reference to a pâté, the term *gratin* simply means that some or all of the meat has been browned in a hot pan before being chilled and ground, adding a rich roasted flavor.

Mousseline Pâtés

Mousseline forcemeats use meat or fish, pureed with cream (fat) and egg whites (the stabilizer that helps to bind and gel them). They are the easiest forcemeat to make at home because they are so stable, the ingredients are commonly available, and you just blend the ingredients in a food processor.

Special Pâtés and Terrines

Any kind of molded preparation that doesn't include traditional forcemeat can be called a terrine. Indeed, such preparations can be every bit as dramatic and enticing as traditional terrines.

PÂTÉ DE CAMPAGNE

A pâté de campagne, or country terrine, is a rustic preparation, slightly more refined than a pâté grandmère (see page 213) mainly in that it uses only a small amount of liver–liver is a seasoning device here rather than the dominant flavor. Also unlike the pâté grandmère, some internal garnish, such as fresh herbs and chunks of smoked ham or duck confit, go a long way. The panade (notice that it's made with flour, not bread) helps to retain moisture and to enrich and bind the pâté.

Most of the meat is ground through a large die, and none of it is pureed, to achieve the charactistic coarse texture of a country terrine. Although only a small amount of liver is used, try to use pork liver if possible rather than chicken liver, because it will allow you to cook the terrine to a lower final temperature and therefore produce a moister pâté.

A pâté de campagne is the easiest terrine to make, and in the spirit of its origins–a humble but delicious dish made from trimmings or inexpensive cuts of meat–should be made with whatever garnish is on hand and eaten simply, with a good baguette and French Dijon. Add a salad of fresh greens, and you've got a simple midweek meal. It's also a fabulous make-ahead dish for a weekend dinner party.

**2 pounds/1 kilogram boneless pork shoulder butt,
cut into 1-inch/2.5-centimeter dice**

4 ounces/100 grams pork or chicken liver

¼ cup/50 grams chopped white or yellow onion

8 tablespoons/48 grams coarsely chopped flat-leaf parsley

1½ tablespoons/24 grams minced garlic

2 tablespoons/25 grams kosher salt

1 teaspoon/3 grams freshly ground black pepper

½ teaspoon/2 grams Pâté Spice (page 145)

2 tablespoons/20 grams all-purpose flour

2 large eggs

2 tablespoons/30 milliliters brandy

½ cup/125 milliliters heavy cream

OPTIONAL GARNISH (MIX AND MATCH TO TASTE)

**Diced ham, cooked mushrooms, rinsed brine-cured green pep-
percorns, duck confit (a total of 1 cup/250
milliliters)**

1. Freeze all your blades and bowls before gathering and measuring your ingredients (see Note below).

2. Preheat the oven to 300 degrees F./150 degrees C.

3. Grind the pork through the large die into the bowl of a standing mixer set in ice. Transfer about one-third of the pork to a smaller bowl, and add the liver, onion, parsley, garlic, salt, pepper, and pâté spice. Fit the grinder with the small die (clean the blade of any sinew that might be caught there) and grind the pork-seasonings mixture into the bowl of coarsely ground pork. Refrigerate.

4. In a small bowl, combine the flour, eggs, brandy, and cream and stir to blend—this is the panade. Add it to the ground meat and, using the paddle attachment, mix until the panade is incorporated and the forcemeat becomes sticky, about a minute. (You can also do this using a wooden spoon or your hands.) Fold in the optional garnish, if using.

5. Do a quenelle test to check the seasoning, and adjust if necessary.

6. Line a 1½-quart/1.5-liter terrine mold with plastic wrap, leaving enough overhang on the two long sides to fold over the top of the terrine when it's filled (moistening the mold first will help the plastic adhere). Fill the mold with the pâté mixture, packing it down to remove air pockets. Fold the plastic wrap over the top, and cover with the lid or with foil.

7. Place the terrine in a high-sided roasting pan and add enough hot water (very hot tap water, 150 to 160 degrees F./65 to 71 degrees C.) to come halfway up the sides of the mold. Put the pan in the oven and bake until the interior of the pâté reaches 150 degrees F./ 65 degrees C. if using pork liver, 160 degrees F./70 degrees C. if using chicken liver, about 1 hour.

8. Remove from the oven, remove the mold from the water bath, and set a weight of about 2 pounds/1 kilogram on top of the terrine. Let cool to room temperature, then refrigerate until completely chilled, overnight, or for up to 1 week, before serving.

Yield: 10 to 12 appetizer servings

[NOTE: See pages 203–210 for a detailed description of the general terrine method.]

PÂTÉ GRANDMÈRE

The kind of pâté a French country grandma would make, very simple, plain, hearty, using plenty of inexpensive, nutrient-rich, intensely flavorful liver. As liver accounts for more than half the meat, it's the dominant flavor. Brian and I prefer to use pork liver rather than chicken livers primarily because you only need to cook pork liver to 150 degrees F./65 degrees C., whereas chicken must reach a temperature of 160 degrees F./ 71 degrees C. So pork liver gives you a moister, and tastier, pâté.

While this is a rustic, coarse, yet soft pâté, attention should be paid to a few steps. Notice that the pork liver is cut into chunks; this increases the surface area of the liver, which will be seared. The more seared surfaces, the better the final flavor. (Chicken livers, which are already small, don't need to be cut up.) Shallots and brandy are then cooked in the same pan and are added to the liver to be ground, for even more flavor.

If you intend to serve the pâté straight from the terrine mold, rather than unmolding it, there's no need to line the terrine. You could even roll the pâté up in plastic wrap and poach in water at 160 to 170 degrees F./71 to 76 degrees C. to the same internal temperature, which will take about an hour. Serve with crusty bread, good Dijon mustard, and sharp, flavored condiments, such as Sweet Pickle Chips (page 300).

1¼ pounds/560 grams pork liver, cut into large chunks,
    or chicken livers
about 2 tablespoons/25 grams kosher salt
1 teaspoon/3 grams freshly ground black pepper
2 bay leaves
2 sprigs fresh thyme
1 pound/450 grams boneless pork shoulder butt, diced
2 tablespoons/30 milliliters vegetable oil
¼ cup/50 grams chopped shallots
2 tablespoons/30 milliliters brandy
2 slices white bread, crusts removed and roughly chopped
½ cup/125 milliliters whole milk
¼ cup/60 milliliters heavy cream
2 large eggs
1 tablespoon/6 grams chopped fresh flat-leaf parsley
¼ teaspoon/2 grams ground white pepper
¼ teaspoon/4 grams freshly grated nutmeg

1. In a large bowl, toss the liver with half the salt and black pepper, 1 bay leaf, and 1 thyme sprig. Toss the pork in another bowl with the remaining salt and pepper, bay leaf, and thyme. Cover and refrigerate (separately) for 8 hours, or overnight.

2. Freeze all your blades and bowls (see Note below).

3. Heat a 12-inch sauté pan over high heat, and add the vegetable oil; the pan should be just smoking. Add the liver without crowding the pieces (discard the bay and thyme), and sauté until it's developed a good sear (the better the "crust," the more flavor). Transfer to a tray and chill in the refrigerator. Add the shallots to the hot pan and sauté until translucent, about 30 seconds, then add the brandy to the pan to deglaze, stirring up any bits of liver and shallot stuck to the bottom. When most of the brandy has cooked off but the shallots still look moist, transfer to a bowl and chill.

4. Preheat the oven to 300 degrees F./150 degrees C.

5. Combine the ingredients for the panada, the bread, milk, cream, and eggs—in a bowl and stir to blend. Set aside.

6. Discard the bay leaf and thyme from the pork. Add the chilled liver and shallots, along with the panada and parsley. Grind this mixture through the small die into the bowl of a standing mixer set in ice.
Careful: The liver is juicy and tends to squirt out of the die.

7. Using the paddle attachment, mix on medium speed for about a minute, until the mixture begins to look sticky.

8. Do a quenelle test to check the seasoning, and adjust if necessary.

9. Line a 1½-quart/1.5-liter terrine mold with plastic wrap, leaving enough overhang on the two long sides to cover the filled mold (moistening the inside of the mold first will help the plastic wrap adhere to the corners). Pack the terrine tightly with the ground mixture: A good way to do this is to use a stiff rubber spatula or large spoon, flicking the spatula quickly down to snap the forcemeat off the spatula into the mold, then jab it down and pack it tightly; this will help prevent air pockets. Fold the plastic wrap over the top of the terrine and cover with the lid or foil.

10. Place the terrine in a high-sided roasting pan and fill the pan with enough hot water (very hot tap water, 150 to 160 degrees F./65 to 71 degrees C.) to come halfway up the sides of the mold. Put the roasting pan in the oven and bake until an instant-read thermometer inserted into the center of the pâté reads 150 degrees F./65 degrees C. if using pork liver, 160 degrees F./71 degrees C. if using chicken liver.

11. Remove the pâté from the oven, remove the terrine from water bath, and set a weight of about 2 pounds/1 kilogram on top of it. Allow to cool to room temperature, then refrigerate until thoroughly chilled, overnight, or for up to a week.

Yield: 24 slices; 12 appetizer servings

[NOTE: See page 203–210 for a detailed description of the general terrine method.]

ENGLISH PORK PIE

Not all classical pâtés en croûte are the highly refined items of the fancy French restaurant. The English have a strong tradition of baking pâtés in a crust. In this rustic preparation, it is a flaky crust. Brian adapted this particular version from my great-grandmother's recipe, made for the family every Christmas morning in Shropshire, England (adaptation was necessary, as Elizabeth Morgan's ground pork contained almost no seasoning). In the family tradition, this was served for breakfast, after Bloody Marys, with Colman's mustard and, of all things, white bread. Brian added garlic and onion, thyme, salt and pepper, and diced ham (which is optional).

You can make aspic the traditional way, using the recipe on page 233, or simply add powdered gelatin to some tasty stock (about 1 tablespoon/10 grams powdered gelatin per 1 cup/250 milliliters liquid will give you a sliceable aspic; see page 233 for working with powdered gelatin). If you don't wish to use lard in the dough, replace it with an equal amount of butter or vegetable shortening.

THE DOUGH
8 ounces/225 grams cold unsalted butter
8 ounces/225 grams cold lard
1 pound/450 grams all-purpose flour
1 large egg
Water as needed

THE MEAT FILLING
1/2 cup/70 grams finely diced onion
1 tablespoon/18 grams minced garlic
1 tablespoon/15 grams unsalted butter
1 ½ pounds/675 grams boneless pork shoulder, diced
1 tablespoon/15 grams kosher salt
1 teaspoon/3 grams freshly ground black pepper
1 teaspoon/3 grams fresh thyme leaves
½ cup/125 milliliters Chicken Stock (page 226), chilled
1 cup/250 grams diced smoked ham (optional)

FOR EGG WASH
1 egg
1 egg yolk
1 tablespoon/15 milliliters whole milk

FOR MEAT TERRINES
Aspic (page 233), as needed (optional)

1. *To make the dough:* Cut the butter and lard into small dice (¼ inch/0.5 centimeter); sprinkle flour on your knife so the butter won't stick. Combine the flour, butter, and lard in a mixing bowl. With your index, and middle fingers and your thumbs, press the fats into the flour until the mixture looks mealy.

2. Crack the egg into a 1-cup/250-milliliter measure, and add enough cold water to equal 1 full cup/250 milliliters. With a fork or whip, beat the water and egg just to combine. Add to the flour mixture and mix just until it forms a paste. Divide the dough in half and shape each piece into a disk. Wrap each one in plastic wrap and refrigerate for at least 1 hour, and up to a day.

3. *To make the meat filling:* Freeze all your blades and bowls before gathering and measuring your ingredients (see Note below).

4. Sauté the onion and garlic in the butter in a medium sauté pan over medium-high heat until soft but not at all browned. Set aside to cool, then chill it in the refrigerator.

5. Combine the meat, salt, pepper, and thyme. Grind through the small die into the bowl at a standing mixer set in ice.

6. Add the cooled onion mixture to the meat and mix with the paddle attachment for 1 minute. Slowly add the chicken stock, mixing until incorporated, about another minute. Add the ham, if using, and mix until incorporated. Cover and refrigerate.

7. *To make the pie:* Preheat the oven to 425 degrees F./220 degrees C.

8. On a floured work surface, roll one piece of dough into a 12-inch/30-centimeter circle about ⅛ inch/0.25 centimeter thick. Brush off the excess flour and place on a baking sheet. Shape the meat mixture into a 5-inch/12.5-centimeter disk about 2 inches/5 centimeters thick and place it in the center of the dough. Carefully lift the edges of the dough and wrap around the meat so that it partially covers the top of the meat. To make the top, roll out the remaining dough ⅛ inch/0.25 centimeter thick and cut out a 6-inch/15-centimeter circle from it. Cut a ¾-inch/2-centimeter hole in the center of the circle. (Reserve dough scraps for another use.)

9. Combine the ingredients for the egg wash in a small bowl and whisk until uniformly blended. Using a pastry brush, paint the edges of the dough encasing the meat, then brush the circle of dough. Lift the top and place egg wash side down on the meat. Crimp the edges of the dough to seal. Brush the entire top with egg wash.

10. Bake the pie for 15 to 20 minutes, until the crust begins to brown. Reduce the oven temperature to 325 degrees F./160 degrees C. and bake until the pie reaches an internal temperature of 150 degrees F./65 degrees C., about 30 minutes longer. Remove the pie from the oven and allow to cool to room temperature, then refrigerate until thoroughly chilled. (You can also serve the pie hot or warm, in which case you would not use aspic.)

11. Warm the aspic until it's just fluid. Pour enough aspic into the steam hole of the pie to reach the top. Refrigerate to set the aspic, about 2 hours.

Yield: 6 servings

[NOTE: See pages 203–210 for a detailed description of the general terrine method.]

PORK TERRINE WITH PORK TENDERLOIN INLAY

Part of the excitement of serving this kind of pâté is that it looks so impressive—a disk of juicy pink tenderloin surrounded by a pâté rich with chunks of garnish. The pork, what is called an inlay in charcuterie terms, should be centered and the dominant feature of the pâté. It entices—to look at it makes you crave it. But although the ground meat for the pâté is pureed in a food processor, rather than mixed, making the texture very fine, and though the tenderloin is seared to enhance flavor, and to help the forcemeat stick to it during cooking, for all its looks and craftsmanship, this pâté is no more difficult to make than a country pâté. (See illustrations on pages 208–209.)

Serve it simply, a slice or two on each small plate with a good condiment (the Onion-Raisin Chutney, for instance, on page 297). It also makes an elegant appetizer, with a sauce (Orange-Ginger, page 292, would work well) or with a salad of mesclun greens with a strong (acidic) vinaigrette.

1 pork tenderloin, about 1 pound/450 grams, fat and silverskin
    removed
1 tablespoon/15 milliliters clarified butter or vegetable oil
1 tablespoon/18 grams minced garlic
1 tablespoon/18 grams minced shallots
1 cup/250 milliliters Madeira
2 tablespoons/30 milliliters brandy
¼ cup/60 milliliters whole milk
2 slices white bread, crust removed
15 ounces/420 grams boneless lean pork shoulder butt, diced
10 ounces/300 grams pork back fat, diced (see step 2 below)
1½ teaspoons/5 grams freshly ground black pepper
½ teaspoon/3 grams pink salt (optional)
2 teaspoons/8 grams Pâté Spice (page 145)

OPTIONAL GARNISH
Diced Ham, black truffle peelings, and/or smoked tongue
    (total of 1 cup/250 milliliters of any or a combination)

1. Freeze all your blades and bowls before gathering and measuring your ingredients (see Note below).

2. Trim the tenderloin to the length of your mold, removing as much of the tapered

end as necessary. The tenderloin will become a central shaft within the terrine, so the more uniform it is, the better; if it's too thick, trim it to the appropriate dimensions relative to your mold to ensure it will be completely surrounded by the forcemeat. Season the tenderloin well with salt and the pepper. Reserve any trimmings and grind with the pork shoulder butt (total weight should still be 10 ounces/280 grams).

3. Place a 10- to 12-inch sauté pan over high heat and add the butter or oil. When it's smoking hot, sear the tenderloin on all sides; the better the sear, the better the flavor. Transfer to a plate lined with a paper towel and put in the refrigerator to cool. Add the garlic and shallots to the pan and sauté briefly, just to soften, about 30 seconds, being careful not to brown them. Add the Madeira and brandy, scraping the bottom of the pan with a wooden spoon to deglaze, and simmer until reduced to a syrupy consistency, a few minutes. Transfer this reduction to a small bowl and refrigerate to chill.

4. Combine the milk and bread, to make the panade, and refrigerate.

5. Preheat the oven to 300 degrees F./150°C.

6. Grind the meat (including any tenderloin trimmings) and fat through the small die into a bowl set in ice. Transfer to the food processor bowl and add the garlic-wine reduction, the panade, pepper, pink salt, if using, and pâté spice. Puree, pulsing, until smooth.

7. If using an optional garnish, transfer the forcemeat to a bowl and fold it in. Do a quenelle test to check for seasoning, and adjust if necessary.

8. Line a 1½-quart/1.5-liter terrine mold with plastic wrap, leaving enough overhang on the two long sides to cover the filled terrine. Fill just under halfway with the forcemeat, spreading it evenly in the mold. Lay the seared tenderloin in the mold, pushing down gently to press the forcemeat up around it, then fill the mold with the remaining forcemeat. Fold the plastic wrap over and cover with the lid or aluminum foil.

9. Place the terrine in a high-sided roasting pan and fill the pan with enough hot water (very hot tap water, 150 to 160 degrees F./65 to 71 degrees C.) to come halfway up the sides of the mold. Put the roasting pan in the oven and bake until an instant-read thermometer inserted into the center of the pâté registers 150 degrees F./60 degrees C.

10. Remove the pâté from the oven, remove from the water bath, and set a weight of about 2 pounds/1 kilogram on top of it. When it's cool enough to handle, place the terrine in the refrigerator and chill overnight, or for up to a week.

Yield: 24 slices; 12 appetizer servings

[NOTE: See pages 203–210 for a detailed description of the general terrine method.]

VENISON TERRINE WITH DRIED CHERRIES

If you have access to venison, this is a delectable dish to make with tougher cuts, such as the shoulder (rather than the loin, for instance). Plus it's an elegant preparation worthy of the winter holidays, especially popular in Eastern Europe, where hunted venison and game are prevalent. Serve it with a spicy-fruity sauce that accentuates the rich flavor of the venison–a chutney, the Orange-Ginger Sauce (page 292), or the traditional Cumberland Sauce (page 291).

> 1 pound/450 grams boneless lean venison shoulder or leg, trimmed of all sinew and cut into 1-inch/2.5-centimeter dice
> 14 ounces/400 grams pork back fat, cut into 1-inch dice
> 1½ cups/375 milliliters dry Madeira
> 1 tablespoon/12 grams Pâté Spice (page 145)
> 2 tablespoons/30 grams kosher salt
> 2 teaspoons/6 grams freshly ground black pepper
> ¼ teaspoon/2 grams pink salt (for color, optional)
> 1 tablespoon/15 milliliters butter
> 1 tablespoon/18 grams minced garlic
> 1 tablespoon/18 grams minced shallots
> 3 large egg whites
> ½ cup/125 milliliters heavy cream
> 2 cups/400 grams dried tart cherries, picked over for pits, plumped overnight in 1 cup/250 milliliters brandy
> 8 to 10 thin (1/8-inch/1/4-centimeter) slices ham

1. Combine the meat, fat, Madeira, pâté spice, salt, pepper, and pink salt, if using, in a bowl or other container, cover, and refrigerate overnight.

2. Freeze all your blades and bowls before gathering and measuring the remaining ingredients (see Note below).

3. The next day, drain the meat, reserving the liquid. Drain the brandy from the plumped cherries and add to reserved liquid. In a sauté pan, melt the butter and sauté garlic and shallots until soft. Add reserved liquid. Bring to a simmer, skimming any scum that rises to the surface. Reduce the liquid to a syrupy consistency, being careful not to burn it. Set aside to cool.

4. Preheat the oven to 300 degrees F./150 degrees C.

5. Grind the meat and fat through the small die into a bowl set in ice. Transfer it to a food processor, add the egg whites, and blend until smooth. Add wine reduction and continue to blend until smooth. Transfer to a chilled mixing bowl. With a rubber spatula, fold in the cream until fully incorporated. Fold in the cherries.

6. Do a quenelle test to check the seasoning and adjust if necessary.

7. Line a 1½-quart/1.5-liter terrine mold with plastic wrap, leaving enough overhang on the two long sides to fold over the top of the terrine. Line with the slices of ham, leaving enough overhang all around to cover the top when the mold is filled. Fill the mold with the forcemeat. Fold the ham and then the plastic wrap over the top, and cover with the lid or aluminum foil.

8. Place the terrine in a high-sided roasting pan and fill the pan with enough hot water (very hot tap water, 150 to 160 degrees F./65 to 71 degrees C.) to come halfway up the sides of the mold. Put the roasting pan in the oven and bake until an instant-read thermometer inserted into the center of the pâté registers 150 degrees F./65 degrees C.

9. Remove the pâté from the oven, remove from the water bath, and set a weight of about 2 pounds/1 kilogram on top of it. Let cool to room temperature, then refrigerate overnight, or for up to 1 week.

Yield: 24 slices; 12 appetizer servings

[N O T E : See pages 203–210 for a detailed description of the general terrine method.]

CHICKEN GALANTINE

Chicken galantine—chicken pâté rolled up inside the skin of the chicken, poached, and served cold—is an exciting and elegant classical presentation, requiring some extra work in the beginning (removing the skin intact from the bird and boning it takes some patience). After that, the preparation follows standard pâté technique, using a straight forcemeat. (See illustrations pages 224–225.)

Part of what gives this preparation its elegance is that it uses the entire bird—the skin becomes the casing, the bones go into the stock the galantine will cook and then cool in to absorb additional flavor. It's best to divide the work into a couple of days: Skin and bone the chicken and make the stock on the first day, then assemble and cook the galantine the second day.

Serve the galantine with Orange-Ginger Sauce (page 292), Cumberland Sauce (page 291), or Onion-Raisin Chutney (page 297).

One 4-pound/2-kilogram chicken, liver reserved

THE FORCEMEAT

Reserved 2 chicken breasts

1 tablespoon/15 grams kosher salt

1 tablespoon/12 grams freshly ground black pepper

2 tablespoons/15 milliliters vegetable oil

1 tablespoon/18 grams minced garlic

1 tablespoon/18 grams minced shallots

1 cup/250 milliliters Madeira

Reserved 10 ounces/280 grams dark meat and trimmings

8 ounces/225 grams pork back fat

Reserved chicken liver

2 large egg whites

½ teaspoon/2 grams Pâté Spice (page 145)

¼ teaspoon/2 grams ground white pepper

¾ cup/185 milliliters heavy cream

OPTIONAL GARNISH (MIX AND MATCH TO TASTE)

½ cup/80 grams unsalted peeled pistachios, roughly chopped

1 cup/100 grams sliced mushrooms, sautéed and chilled

¼ cup/24 grams chopped fresh herbs (parsley, tarragon, chives, and/or chervil)

¼ cup/50 grams black truffle peelings

¼ cup dried cherries

IF COOKING THE GALANTINE USING THE TRADITIONAL METHOD

Cheesecloth for rolling

About 2 quarts/2 liters Chicken Stock (recipe follows), or as needed

IF POACHING THE GALANTINE IN WATER

Plastic wrap

1. *To remove the skin and bone the bird:* Sharpen your boning knife. Remove the wings at the second joint, leaving the drumettes attached to the carcass. Make a cut around the base of each leg, cutting through the skin to the bone. Turn the bird over, breast side down, and make an incision down the back from the neck to the tail, cutting only through the skin. Working down the sides, start to remove the skin by sliding your knife between skin and flesh; be careful not to tear or poke holes in the skin. When you reach the base of the legs and drumettes, carefully pull them through the skin, as if you were removing a tight shirt. Continue passing your knife between skin and flesh as you work your way over the breast until you can remove the entire skin. You will have a piece of skin about 1 foot/30 centimeters long and 8 inches/20 centimeters wide, with four holes in it from the legs and wings. Trim it to a neat rectangle about 8 inches/20 centimeters by 12 inches/30 centimeters, or as large as you're able to make it.

2. Cover a baking sheet with plastic wrap (it helps to moisten the pan first so the plastic will stick). Lay the chicken skin on it, outside down, arranging it so there are no wrinkles. Freeze for 1 hour.

3. Meanwhile, remove the legs and thighs from the chicken, then take the meat off the bone, making sure to remove all fat and sinew. Remove the two breast halves, then remove the tenderloin from each one and remove the sinew that runs throughout it. (Reserve all the bones, the wings, and skin for the stock.) Trim the breasts to about 3 inches/7.5 centimeters by 5 inches/12.5 centimeters–they will become the inlay, and the trimmings and the leg and thigh meat will be used in the forcemeat. You should have about 10 ounces/280 grams of cleaned dark meat and trimmings from the breasts. Cover and refrigerate all the meats until you need them.

4. Remove the frozen skin from the freezer and, with a sharp knife scrape away excess fat pockets. The skin should become pliable within 4 to 5 minutes.

5. *To make the forcemeat:* Freeze all your blades and bowls before gathering and measuring your ingredients (see Note below).

6. Season the chicken breasts aggressively with salt and pepper (remembering it will be served cold). Heat the butter or oil in a small sauté pan over high heat. When it's smoking, add the breasts and sauté until nicely browned on both sides but still raw in the center. Remove and set aside to cool.

7. Add the garlic and shallots to the pan and cook until translucent, about 1 minute. Deglaze with the Madeira, scraping up the browned bits on the bottom of the pan with a

THE GALANTINE

A chicken galantine is a special pâté that in effect trades the skin of the bird for the terrine mold. Here, a sheet of plastic wrap has been laid down on the work surface and a layer of cheesecloth spread over it. The skin from the chicken, which was removed in one piece, scraped of excess fat, and squared off into a rectangle, is laid on the cheesecloth. The pâté forcemeat, with its secondary garnish (truffles, pistachios, and dried cherries), is spread out on the skin. Whole pieces of chicken breast are laid down the center of the forcemeat.

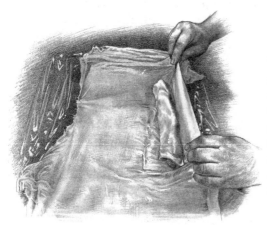

The plastic wrap and cheesecloth are used to help roll the skin around the pâté, forming a tight cylinder, with the chicken breast in the center.

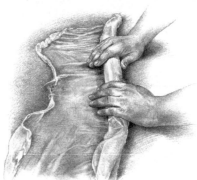

The galantine is rolled tightly to ensure a good, even texture.

wooden spoon, then reduce until the mixture is almost a paste. Transfer the reduction to a small bowl, cover and refrigerate.

8. Combine the dark meat, fat, and liver and grind through the small die. Transfer to the bowl of a food processor, add the egg whites, salt, pepper, pâté spice, and shallot-Madeira reduction, and puree until smooth, about 3 minutes. Transfer to a bowl set in ice, and, using a rubber spatula, blend in the heavy cream.

9. Do a quenelle test to check for seasoning, and adjust the seasoning if necessary. Fold in the optional garnish, if using. Refrigerate until chilled.

10. *To cook the galantine using the traditional method:* Cut a double thickness of cheesecloth large enough to roll the galantine in, about 12 inches/30 centimeters by 2 feet/60 centimeters. Lay the skin outside down in the center. Place half the forcemeat down the center in a rectangular shape about 4 inches/10 centimeters by 8 inches/20 centimeters. Lay the seared

The galantine will be completely wrapped in cheesecloth, tied off at each end, and given further support by strips of cheesecloth tied around its girth.

The galantine is cooked in chicken stock that is just below a simmer, about 160 to 180 degrees F./71 to 82 degrees C. A hotel pan is being used here, but a standard pot works fine so long as the galantine is completely submerged. A fish poacher is actually the best cooking vessel, in that its shape requires less stock. When the galantine has reached an internal temperature of 160 degrees F./71 degrees C., the pan is removed from the stovetop and the galantine is cooled and chilled submerged in the flavorful stock.

After the galantine has been chilled overnight, it is removed from the gelatinous stock and unwrapped. It is ready to be sliced.

breasts end to end down the center and cover with the remaining forcemeat. Fold the skin up over so both sides meet and you have a compact roulade about 3 inch/7.5 centimeter in diameter. Roll up in the cheesecloth as tightly as possible, and twist and tie each end with string. Cut three strips of cheesecloth ½ inch/1 centimeter wide by about 12 inches/30 centimeters long and tie around the galantine for additional support.

Heat the chicken stock in a pan large enough to hold the galantine (a fish poacher is the perfect shape) to a temperature of 170 degrees F./76 degrees C. Poach the galantine in the stock to an internal temperature of 160 degrees F./71 degrees C., 30 to 45 minutes, weighting it down with a rack or plate to keep the galantine submerged. Remove from the heat and let the galantine cool to room temperature, then refrigerate it, still in the poaching liquid.

11. *To cook the galantine in water:* Form and roll the galantine as directed above, substituting plastic wrap for cheesecloth. Tie one end with butcher's string. Holding on to the knot, roll the galantine along the counter to tighten it. Twist the other end of the plastic wrap and roll the galantine along the counter to tighten it, then tie that end with string. The galantine should be nice and tight. Repeat the process with a second layer of plastic wrap for additional strength.

Follow the cooking instructions in step 10 above, using water instead of stock. Let cool in the water to maintain its shape, then remove and refrigerate.

12. To serve, remove the cheesecloth or plastic wrap and slice the galantine.

Yield: 16 to 20 slices; 8 to 10 appetizer servings

[NOTE: See pages 203–210 for a detailed description of the general terrine method.]

Chicken Stock

This is a basic recipe for chicken stock. Because the bones are not roasted, they'll release a lot of fat and protein that will collect on the surface so it's important to skim the stock diligently when it first comes to a boil. It should then be cooked at the lowest possible temperature, the surface just at a tremble–this will result in a very clear, clean-tasting stock. If a stock is boiled, it will become cloudy as the fat and other impurities are emulsified into it.

1 cup/150 grams roughly chopped onions
½ cup/75 grams roughly chopped celery
½ cup/75 grams roughly chopped carrots
2 bay leaves
1 bunch fresh thyme
10 black peppercorns, crushed with the side of a knife
2 cloves garlic
8 pounds/4 kilograms chicken bones, necks, feet, and giblets
6 quarts/6 liters cold water

1. Combine all the ingredients except the water in a pot that is taller than it is wide. Add the water; it should cover the ingredients by about 1 inch/2.5 centimeters. Slowly bring the water to a boil then turn to the lowest possible simmer. Skim the stock to remove the impurities on surface. Gently simmer for 5 hours, skimming occasionally.

2. Strain the stock through a fine-mesh strainer. Cool completely before refrigerating.

Yield: 1 gallon/4 liters

ROASTED DUCK ROULADE

This is an excellent alternative to traditional roast duck. The flavors are exciting, and the slice-and-serve preparation is stunning; and prepared this way, one duck makes many more servings than if roasted whole. The hard part of this roulade, as with any galantine-type preparation, is removing the skin intact. (Once the skin is trimmed, you should have a rectangle about 10 inches/25 centimeters long and 5 inches/12.5 centimeters wide.) Scraping the fat off the frozen skin also takes some time and care. After that, though, this is prepared like any other pâté, then rolled up in the skin and gently roasted with aromatic vegetables. You might want to roast the bones to make a stock or the base for a sauce or jus.

This can be served with the traditional accompaniments to roast duck. We suggest serving it on a bed of French lentils, along with some haricots verts (fine green beans).

1 Pekin (Long Island) duck, about 4 pounds/2 kilograms,
    skinned, boned, and sinews removed (see headnote
    above and Chicken Galantine, page 221), liver reserved
4 ounces/112 grams pork back fat (or as needed), diced
1 tablespoon/15 grams kosher salt
1 ½ tablespoons/5 grams white pepper
1 tablespoon/15 milliliters vegetable oil
1 tablespoon/18 grams minced shallots
1 cup/250 milliliters dry sherry
1 tablespoon/6 grams minced fresh sage
2 tablespoons/30 grams Roasted Garlic Paste (page 125)

FOR COOKING THE ROULADE
8 ounces/280 grams unsalted butter
1 cup/250 grams diced onion
1 cup/250 grams chopped carrots
1 ½ tablespoons/30 grams minced garlic

2 bay leaves

1 bunch fresh thyme

Kosher salt and freshly ground black pepper

Parchment paper

1. *Weigh the dark meat:* You should have about 12 ounces/336 grams. Add enough pork fat to equal a total of 1 pound/450 grams. Refrigerate. Cut the duck breasts into large dice, and refrigerate.

2. Lay the duck skin on a plastic wrap–lined baking sheet, outside down, arranging it so there are no wrinkles. Freeze for 1 hour.

3. Freeze all your blades and bowls before gathering and measuring the remaining ingredients (see Note below).

4. Season the duck breast aggressively with salt and pepper. Heat the oil in a medium sauté pan over high heat. When it's almost smoking, add the diced breast and sauté until the pieces are nicely browned on all sides but still raw in the center. Remove and set aside to cool.

5. Add the shallots to the pan and cook until translucent, about 1 minute. Deglaze with the sherry, scraping up the browned bits from the bottom of the pan with a wooden spoon, then reduce until the mixture is almost a paste. Transfer the reduction to a small bowl and refrigerate it until chilled.

6. Combine the duck leg and thigh meat and fat, along with the liver and grind through the small die into the bowl of a standing mixer bowl set in ice. Add the sage, roasted garlic, salt and pepper, and the chilled reduction and mix on low speed, using the paddle attachment, for 1 to 2 minutes, just until well combined (don't overmix, or the fat will become too hot). Fold in duck breast. Refrigerate.

7. Do a quenelle test to check the seasoning, and adjust if necessary. Cover and refrigerate.

8. Remove the skin from freezer and scrape off the excess fat (there will be a lot), being careful not to cut or tear the skin.

9. Place the duck mixture down the middle of the skin and roll up into a log or roulade. Tie each end securely with butcher's twine, making sure to pinch the skin to trap the filling. Tie one loop of string lengthwise around the roulade making it just snug. Tie individual loops of string around it, as with a roast, to make a tight, uniform roulade or log. Season with salt and pepper.

10. Preheat the oven to 325 degrees F./160 degrees C.

11. In a large heavy skillet, melt the butter over medium-high heat. Add the onion, carrots, and garlic, along with the herbs and salt and pepper to taste, and sauté until they become translucent.

12. Transfer the vegetables and cooking juices to a baking sheet or roasting pan and spread out to make a bed the width and length of the roulade (the aromatic vegetables and herbs will flavor the basting liquid and thus the roulade). Place the roulade on top, and cover loosely with buttered or oiled parchment paper (this deflects the heat somewhat, making it a little more gentle, and also inhibits moisture loss).

13. Bake, basting the roulade frequently with the butter and juices in the pan, until an instant-read thermometer inserted in the center of the roulade reaches 140 degrees F./60 degrees C., about 45 minutes to 1 hour.

14. Remove the parchment paper and raise the oven temperature to 375 degrees F./190 degrees C. to brown the skin. Continue roasting until the internal temperature reaches 150 degrees F./65 degrees C., about 15 more minutes. Let the roulade rest for 15 minutes before slicing into ½-inch/1-centimeter to 1-inch/2.5-centimeter slices.

Yield: 8 slices; 4 main-course servings

[NOTE: See pages 203–210 for a detailed description of the general terrine method.]

PORK PÂTÉ EN CROÛTE

The term *en croûte* (in a crust) refers to an ingredient encased in pastry. This is the most difficult and most luxurious way to use forcemeat, the apotheosis of the charcutier's craft. The trick is in browning the dough and cooking it all the way through without over- or undercooking the forcemeat inside (this dough originated from work by Dan Hugelier and Lyde Buchtenkirch for the Culinary Olympics team; to enhance the browning, they introduced milk powder). When you succeed, it's a true victory. And, involved though it is, none of the steps are difficult in and of themselves (you're baking meat loaf in a piecrust).

This terrine uses the pork terrine on page 218, but you can use any farce and inlay you wish. You will need a pâté en croûte mold–12 by 3 by 3 inches/30 by 7.5 by 7.5 centimeters–with a removable bottom (see Sources, page 306).

PÂTÉ EN CROÛTE

A pâté en croûte is a pâté baked in a pastry crust. Here the supple dough is rolled into a rectangle about ¹⁄₁₆ inch/0.125 centimeter thick.

To determine the desired dimensions of the dough, the mold (the bottom of the mold is removable) is pressed into the rolled dough, marking the dimensions of its sides and ends.

The excess dough is cut away and reserved for later; the dough should extend at least ½ inch/ 1 centimeter beyond the mold outline (any excess can always be trimmed away).

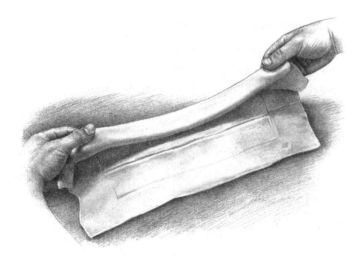

The dough is then gently folded accordion-style, or into a ribbon fold, so it can be lifted up and then unfolded in the mold.

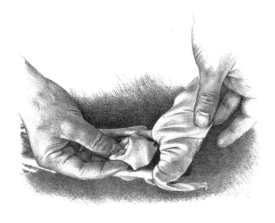

A floured dough ball is used to press the dough into all the corners and angles of the mold.

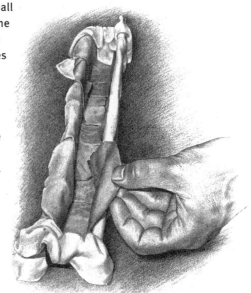

Thin slices of ham are laid into the mold, slightly overlapping. The main function of the ham is as a moisture barrier between the forcemeat and dough, to prevent the dough from becoming soggy. (The ham will also form an elegant border in the sliced pâté and add flavor.) Forcemeat is next added to fill the mold halfway, just as with a pâté en terrine. The main garnish, a whole seared pork tenderloin, is laid in, then covered with the remaining forcemeat. The ham overhanging the sides of the mold is folded over the top.

The pâté en croûte is sealed by folding the dough over the ham, brushing it with egg wash to help it adhere. This, the only seam in the dough, will become the bottom of the pâté.

The bottom of the mold has been replaced and the mold flipped back over. The top of the pâté dough is given a final brushing of egg wash for color. (With a terrine this small, there is no need to make a steam hole; the forcemeat will cook before the dough can become soggy.)

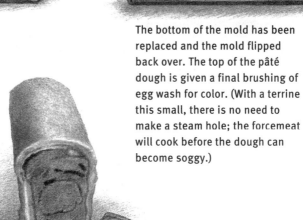

A slice of the finished pâté en croûte: The pâté fills the interior of the crust (the fat has not broken out), the tenderloin is perfectly centered, and the pâté is loaded with colorful and flavorful random garnish.

Pâté Dough (recipe follows)
8 paper-thin slices smoked ham (about 4 inches/
10 centimeters by 6 inches/15 centimeters)
Pork Terrine with Pork Tenderloin Inlay (page 218), prepared
through step 7
1 large egg, beaten for egg wash
Aspic for Meat Terrines (page 234), as needed

1. Preheat the oven to 450 degrees F./230 degrees C. Have ready a hinged en croûte mold (12 by 3 by 3 inches/30 by 7.5 by 7.5 centimeters; see Note below) with a removable bottom.

2. On a floured surface, roll the pâté dough out to an 12-by-20-inch/30-by-50-centimeter rectangle, about $\frac{1}{16}$ inch/0.125 centimeter thick.

3. Remove the bottom of the mold and set it aside. Place the mold upside down on a baking sheet. Gently fit the dough into the mold, being careful not to tear it. Pinch off a small piece of the excess dough, ball it up, dust it with flour, and use it to press the dough into the corners of the mold.

4. Line the dough with the slices of ham, slightly overlapping them and leaving enough overhang to cover the filled terrine.

5. Fill the mold with half of the forcemeat. Lay the tenderloin in the mold, then, pushing it down gently, fill the mold with the remaining forcemeat. Fold the ham over the top. Trim the dough so it will overlap by at least $\frac{1}{2}$ inch/1 centimeter, and fold it over the top, completely enclosing the pâté, and brushing the seam with egg wash to seal. Slide the bottom of the mold onto what is now the top, then invert the mold onto another baking sheet. Brush egg wash over the top, being careful not to allow egg to seep down into mold (where it could stick and then tear your dough when you unmold it).

6. Bake the pâté until the dough is nicely browned, about 20 minutes. Remove the pâté and set on a rack to rest while the oven cools. Reduce the oven temperature to 325 degrees F./160 degrees C. and bake until an instant-read thermometer inserted in the center reaches 150 degrees F./65 degrees C. Cover the crust loosely with foil if it starts to brown too much. Let cool to room temperature.

7. Refrigerate for 12 to 24 hours before unmolding and slicing.

Yield: 24 slices; 12 appetizer servings

[NOTE: For larger terrine molds, you may want to cut a steam hole in the crust to prevent the crust from becoming soggy. See steps 11 and 12 on page 242.]

Pâté Dough

8½ ounces/235 grams bread flour
¾ ounce/17 grams nonfat dry milk powder
1½ teaspoons/11 grams kosher salt
3 ounces/84 grams unsalted butter, softened
1 large egg
5 tablespoons/75 milliliters whole milk
1 teaspoon/5 milliliters white or cider vinegar

1. Combine the flour, milk powder, and salt in the bowl of a standing mixer fitted with a dough hook, and mix to combine. Add the butter bit by bit.

2. Combine the egg and milk in a small bowl, then add the vinegar. Add to the flour mixture and mix until a stiff dough forms.

3. Shape into a disk, wrap in plastic, and refrigerate for at least 1 hour, and up to 1 day before using.

Yield: 10 ounces/300 grams dough

Aspic

Aspic is nothing more than flavorful stock that's been clarified and then set with gelatin—in other words, jelled consommé. For very rustic preparations, such as the English Pork Pie (page 215), you don't even need to clarify it—though traditionally you wouldn't use powdered gelatin either. But clarifying the aspic makes a visual impression worthy of the effort, especially if you add fresh thyme leaves to it or other chopped herbs appropriate to the dish. And the work of clarifying stock is easy and satisfying.

Using powdered gelatin to make aspic: The two stages involved in working with powdered gelatin are blooming and melting. First, sprinkle the gelatin over 3 times its volume in water in a baking dish and allow it to "bloom": the gelatin will absorb the water and turn translucent. Next, place the bloomed gelatin in a 350-degree-F./175-degree-C. oven for 5 minutes, or until it is melted. It's now ready to be added to the liquid being gelled.

Half an ounce/14 grams (about 1 tablespoon) of powdered gelatin per 1 cup/250 milliliters liquid will give you a sliceable aspic. One teaspoon/4 grams per 1 cup/250 milliliters results in a more delicate gel.

Aspic for Meat Terrines

4 ounces/112 grams lean ground chicken or turkey

½ cup/70 grams finely chopped onion

¼ cup/35 grams finely chopped carrot

1 Roma (plum) tomato, chopped

2 large egg whites

1 cup/250 milliliters dry Madeira

1 bay leaf

1 quart/1 liter rich chicken stock or White Veal Stock (page 269)

3 ounces/90 milliliters cold water

1 ounce/25 grams unflavored powdered gelatin
(2 tablespoons)

1. In a large heavy-bottomed nonreactive saucepan, combine the meat, onion, carrot, tomato, egg whites, and Madeira and whisk to mix well. Add the stock and slowly bring to a simmer, stirring continuously, so the egg white doesn't stick to the bottom and burn, until a raft forms. Once the raft forms, stop stirring and gently simmer the stock for 1 hour; do not allow to come to a boil.

2. Carefully strain the stock through a dampened coffee filter: It should be crystal clear. Let cool to room temperature.

3. Place the cold water in a dish, sprinkle the gelatin over, and allow it to absorb the water (bloom).

4. Heat half the clarified stock in a small saucepan and add the bloomed gelatin. Simmer briefly until it dissolves.

5. Remove from the heat, place a few tablespoons on a plate, and refrigerate to check the firmness. Once chilled, the aspic should be firm enough to cut but not so firm that it would bounce like a rubber ball.

Yield: 3 cups/750 milliliters

VEAL TERRINE GRATIN

This is an elegant classical terrine, using veal, with pork shoulder and a little bit of bacon that have been seared. (Unlike the classic pâté grandmère and pâté de campagne, though, this one contains no liver.) The searing adds a depth of flavor, a more roasted flavor. Garlic, shallots, Madeira, and brandy, used in deglazing the pan used to sear the pork shoulder, also add sweetness and depth to the flavor. (If you happen to have some *glace de viande*–reduced meat stock–on hand, it can be added to the reduction for even more meaty taste.) Adding green pistachios and chopped black truffle results in an exciting visual appearance and texture.

Because the veal will be ground, buy an inexpensive cut, shoulder, leg, or any stewing veal. Serve this with the classic Cumberland Sauce (page 291).

5 ounces/140 grams boneless pork shoulder butt, diced

1 tablespoon/20 grams kosher salt

1½ teaspoons/5 grams freshly ground black pepper

1 tablespoon/15 milliliters vegetable oil

1 ounce/25 grams slab bacon, diced

1 tablespoon/18 grams minced garlic

1 tablespoon/18 grams minced shallots

1 cup/250 milliliters dry Madeira

2 tablespoons/30 milliliters brandy

10 ounces/280 grams boneless lean veal, diced

10 ounces/280 grams pork back fat, diced

2 slices white bread, crust removed

¼ cup/60 milliliters whole milk

1 large egg

½ teaspoon/3 grams pink salt (optional)

2 teaspoons/8 grams Pâté Spice (page 145)

OPTIONAL GARNISH (MIX AND MATCH TO TASTE)

1½ cups diced smoked ham; pork confit or duck confit;
    mushrooms, raw or sautéed; chopped fresh herbs,
    such as parsley or chives

1. Freeze all your blades and bowls before gathering and measuring your ingredients (see Note below).

2. Season the pork shoulder with salt and pepper. Place a 10- to 12-inch sauté pan over high heat. Add the butter or oil. When it's smoking hot, add the pork shoulder and bacon and sear on all sides (the better the sear, the better the flavor). Transfer to a plate lined with a paper towel and place in the refrigerator to cool.

3. Add the garlic and shallots to the pan and sauté briefly just to soften, about 30 seconds, being careful not to brown them. Add the Madeira and brandy, scraping up the browned bits from the bottom of the pan with a wooden spoon, to deglaze and simmer until reduced to a syrupy consistency, a few minutes. Transfer this reduction to a small bowl and chill in the refrigerator.

4. Preheat the oven to 300 degrees F./150 degrees C.

5. Grind the veal, seared pork, and back fat through the fine die into a bowl set in ice. Transfer to the food processor bowl. Combine the bread and milk, then add to the processor, along with the chilled garlic-Madeira reduction, the egg, the salt, the pink salt, if using, and pâté spice, and pulse until smooth.

6. Transfer the meat mixture to a bowl. If using the optional garnish, set the bowl in ice, then fold in the garnish. Do a quenelle test to check the seasoning (keep the remaining mixture refrigerated while you do so) and adjust the seasoning as needed.

7. Moisten a terrine mold with water (to anchor the plastic wrap) and line with plastic wrap, leaving enough overhang on the two long sides to fold over the top. Pack the forcemeat tightly into the mold, pressing it down to get rid of any air pockets. Fold the plastic wrap over the top, then cover the terrine with the lid or foil.

8. Place the terrine in a high-sided roasting pan and add enough hot water (very hot tap water, 150 to 160 degrees F./65 to 71 degrees C.) to come halfway up the sides of the mold. Put the roasting pan in the oven and bake until an instant-read thermometer inserted into the center of the pâté registers 150 degrees F./65 degrees C. Remove the pâté from the oven, remove from the water bath, and set a weight of about 2 pounds/1 kilogram on top of it. Refrigerate overnight.

Yield: 24 slices; 12 appetizer servings

[NOTE: See pages 203–210 for a detailed description of the general terrine method.]

SHRIMP AND SALMON TERRINE WITH SPINACH AND MUSHROOMS

This is probably the simplest terrine to make at home. It has a rich, smooth shrimp and salmon flavor, and the garnish–a mosaic of green spinach surrounding the salmon fillet in the center–is visually dramatic. The recipe calls for small shrimp (26 to 30 per pound), because they tend to be the least expensive, but any shrimp will work well. This would also work well, both visually and from a flavor standpoint, with Basil Cream Sauce (page 293).

1 tablespoon/15 milliliters vegetable oil

8 ounces/225 grams mushrooms, trimmed and sliced
¼ inch/0.5 centimeter thick

1 tablespoon/15 grams kosher salt

4 ounces/110 grams spinach leaves (about 2 cups), stems
removed and cut into chiffonade (slivered)

1 pound/450 grams small (26/30 count) shrimp, peeled and
deveined or rock shrimp

1 large egg white

1 cup/225 milliliters heavy cream

Pinch of white pepper

10 ounces/280 grams skinless salmon fillet: if trimming a side
of salmon specifically for this recipe (see Note 1 below),
you want a long piece 1 to 1½ inches/2.5 to 3.5
centimeters wide by 1 inch/2.5 centimeters thick and
the length of the mold (see step 6 below); or buy two
5-ounce/140-gram center-cut pieces salmon fillet

1. Freeze all your blades and bowls before gathering and measuring your ingredients (see Note 2 below).

2. Heat a large sauté pan, then add the butter or oil and heat until almost smoking. Add the mushrooms, season with 1 teaspoon/7 grams salt, and sauté until soft. Remove from the pan and chill.

3. Preheat the oven to 300 degrees F./150 degrees C.

4. Place the shrimp and egg white in the food processor and puree. While the machine is running, add the cream in a steady slow stream. Season with the remaining salt and the white pepper and process to smooth paste. (For an exquisite texture, pass this mixture through a drum sieve, or tamis.)

5. Transfer the shrimp mixture to a bowl and fold in the cooled mushrooms and spinach.

6. Line a 1½-quart/1.5-liter terrine mold with plastic wrap, leaving enough overhang on the two long sides to cover the filled mold (moistening the mold first will help the plastic wrap adhere to the corners). Fill with two-thirds of the shrimp mixture, lay the salmon fillet in the center (if using 2 fillets, lay them end to end, trimming them as necessary to fit) and push down with your hands so the mixture mousse comes up around the sides of the salmon. Top with the remaining shrimp mixture and fold the plastic wrap over to cover. Cover with the lid or foil.

7. Place the terrine in a high-sided roasting pan and add enough hot water (very hot tap water, 150 to 160 degrees F./65 to 71 degrees C.) to the pan to come halfway up the sides of the mold. Put the roasting pan in the oven and bake until an instant-read thermometer inserted into the center of the pâté registers 140 degrees F./60 degrees C.

8. Remove the pâté from the oven, remove from the water bath, and set a weight of 2 pounds/1 kilogram on top of it. Allow to cool, then refrigerate.

9. To serve, remove the pâté from the mold and pat dry with paper towels. Cut into ¼-inch/0.5-centimeter-thick slices.

Yield: 24 slices; 12 appetizer servings

[NOTES: 1. If you're cutting the salmon yourself, you'll have a lot of trimmings. Use them for or add them to a seafood sausage (page 146) or salmon pâté (page 240), or cure and hot-smoke (see pages 52 and 74) for a smoked salmon spread, mixing to taste with some butter, crème fraîche, sour cream, lemon juice, and fresh herbs. 2. See pages 203–210 for a detailed description of the general terrine method.]

MARYLAND CRAB, SCALLOP, AND SAFFRON TERRINE

An Eastern Seaboard terrine–a scallop mousseline flavored and colored a vibrant yellow with saffron, and garnished with whole chunks of crab and green chives. This would be enhanced accompanied by a roasted red pepper aïoli, a gribiche sauce, a cucumber dill salsa, or rémoulade sauce, and it would even go well with the smoky tomato salsa.

8 leeks, green tops only (whites reserved for another use, such
as soup or sauce, or simply braise them in stock)

¾ cup/185 milliliters heavy cream

A large pinch of saffron threads

1 pound/450 grams sea scallops

2 large egg whites

1 tablespoon/15 grams kosher salt

1 teaspoon/3 grams ground white pepper

1 teaspoon/5 milliliters fresh lemon juice

1 pound/450 grams Maryland lump or jumbo lump crabmeat,
picked over for shells and cartilage

¾ cup/100 grams chopped fresh chives

1. Freeze all your blades and bowls before gathering and measuring your ingredients (see Note below).

2. Wash the leek greens thoroughly, and split them lengthwise so you end up with long strips about 2 by 8 inches/5 by 20 centimeters. Cook them in a large pot of heavily salted boiling water until completely tender, about 8 minutes. Drain and chill in ice water, then drain and pat dry. Lay out on a sheet of plastic wrap or wax paper.

3. In a small saucepan, bring the cream to a boil over high heat; remove from the heat. Add the saffron and let it sit for 15 minutes to infuse the cream, then chill, uncovered, in the refrigerator.

4. Preheat the oven to 300 degrees F./150 degrees C.

5. Combine scallops with the egg whites in a food processor and puree until smooth. While the machine is running, add the saffron cream in a slow, steady stream. Season with the salt, pepper, and lemon juice. Transfer to a bowl and fold in the crabmeat and chives. Cover with plastic wrap and refrigerate.

6. Line a 1½-quart/1.5-liter terrine mold with plastic wrap, leaving enough overhang on the two long sides to cover the filled terrine (moistening the mold first will help the plastic wrap adhere to the corners). Then line it with the leek greens, laying them crosswise in the mold and leaving enough overhang to cover the top. Pack the scallop mixture into the mold. Fold the leek greens over the top, followed by the plastic wrap. Cover with the lid or aluminum foil.

7. Place the terrine in a high sided roasting pan and add enough hot water (very hot tap water, 150 to 160 degrees F./65 to 71 degrees C.) to the pan to come halfway up the side of the mold. Bake until an instant-read thermometer inserted in the center of the pâté registers 140 degrees F./60 degrees C.

8. Remove the pâté from the oven, remove from the water bath, and set a weight of about 2 pounds/1 kilogram on top of it. Let cool, then refrigerate overnight.

Yield: 24 slices; 12 appetizer servings

[NOTE: See pages 203–210 for a detailed description of the general terrine method.]

SALMON PÂTÉ IN BASIL CORNMEAL CRUST

This is a highly refined pâté en croûte, an elegant version of other well-known dishes in which fish is cooked in a pastry crust, such as the Russian coulibiac. It uses a salmon forcemeat, with shrimp, salmon, and fresh basil and chives as the bright flavorful interior garnish. The herbed cornmeal crust makes this a very special pâté en croûte. Serve it with Cucumber Dill Relish (page 283), Smoked Tomato and Corn Salsa (page 284), Rémoulade (page 281), or Sauce Gribiche (page 282).

This, like all classical pâtés en croûte, requires a special mold–12 by 3 by 3 inches/30 by 7.5 by 7.5 centimeters–with a removable bottom (see Sources, page 306). There are many steps involved–making the dough, making the forcemeat, cooking and cooling the pâté, preparing the aspic–so it's a good idea to read the recipe through first to decide on a good plan for yourself.

THE DOUGH

about 1 cup/168 grams cornmeal

about 2 cups/308 grams bread flour

1 ½ tablespoons/20 grams kosher salt

¼ cup/24 grams firmly packed chopped fresh basil

¼ cup/24 grams chopped fresh flat-leaf parsley

2 large eggs

½ cup/125 milliliters whole milk

1 tablespoon/15 milliliters white vinegar

THE SALMON FORCEMEAT

1 pound/450 grams salmon fillet, diced (reserve one strip, as noted in the garnish ingredients)

2 large egg whites

1 tablespoon/15 grams kosher salt

1 teaspoon/3 grams ground white pepper

1 cup/250 milliliters heavy cream

THE GARNISH

12 ounces/330 grams (21/25 count) shrimp, peeled and
     deveined

¼ cup/24 grams chopped fresh basil

2 tablespoons/16 grams chopped fresh chives

10 ounces/280 grams skinless salmon fillet: if trimming a
     side of salmon specifically for this recipe (see Note 1
     below), you want a long piece 1 to 1½ inches/2.5 to 3.5
     centimeters wide by 1 inch/2.5 centimeters thick and
     the length of the mold (see step 6 below); or buy two
     5-ounce/150-gram center-cut pieces salmon fillet

Aspic for Seafood Terrines (page 243) as needed

1 egg, beaten, for egg wash

1. *To make the dough:* Sift the dry ingredients together and place in a food processor. Add the basil and parsley and process until the mixture takes on a uniform green hue. Add the eggs, milk, and vinegar and pulse until the dough starts to come together, about 10 times.

2. Dump the mixture out onto a floured surface and knead until a smooth, elastic dough forms, 3 to 5 minutes. Pat it down into a rectangle about 2 inches/5 centimeters thick, wrap in plastic, and refrigerate for 1 hour.

3. *To make the forcemeat:* Freeze all your blades and bowls before gathering the remaining ingredients (see Note 2 below).

4. Combine the diced salmon, egg whites, salt, and pepper in a food processor and puree until smooth, 1 to 2 minutes. While the machine is running, slowly add the cream, stopping to scrape down the sides of the bowl midway through mixing. Transfer to a medium bowl.

5. Do a quenelle test (see page 136) to check for seasoning (keep the remaining mixture refrigerated while you do so), and adjust the seasoning if needed. Fold in the shrimp, basil, and chives. Cover and refrigerate.

6. Preheat the oven to 400 degrees F./200 degrees C. Have ready a hinged en croûte mold (12 by 3 by 3 inches/30 by 7.5 by 7.5 centimeters) with a removable bottom.

7. On a floured surface, roll the dough out to a rectangle about 12 by 20 inches/30 by 50 centimeters, and about 1/16 inch/0.125 centimeter thick. Remove the bottom of the mold and spray it and the insides with vegetable oil; set the bottom aside. Place the mold in the center of the rolled dough and, one at a time, roll it over onto each side on the dough, making a slight indentation where the top and sides hit the dough each time to mark it (see illustrations on pages 230–231).

8. Following the impressions in the dough, cut out a rectangle that is roughly 1½ inches/3.5 centimeters longer and wider than the mold; reserve the scraps. Place the mold on a baking sheet pan and line it with the dough: Fold the bottom third of the dough up, as if folding a letter, then fold the top third under (called a ribbon fold, this will allow you to place the dough in the mold without its tearing), and gently fit it into the mold. There should be an overhang of 1½ inches/3.5 centimeters all the way around. Press the dough into all the corners, being careful not to tear it; if that happens, repair it by pressing the dough together or patching it with some of the reserved dough.

9. Fill the mold with half of the forcemeat. Lay the salmon fillet in the center and press down so the forcemeat oozes up all around it, then fill the mold with the remaining forcemeat. Trim both ends of the overhanging dough to ½ inch/1 centimeter, then fold them over. Fold one long side of the dough over the salmon; it should cover it about halfway. Brush the dough with some egg wash. Then fold the other long side over and press so the dough sticks together. Slide the bottom of the mold onto what is now the top, then invert the mold onto another baking sheet. Brush egg wash over the top, being careful not to let any drip down into the mold (it could stick when cooked and then tear your dough when you unmold the pâté).

10. Bake the pâté for 20 minutes. Remove the pâté and set on a rack to rest while the oven cools. Reduce the oven temperature to 325 degrees F./160 degrees C.

11. Cut a hole in the center of the top of the pâté about ¾ inch/2 centimeters in diameter. Using a piece of foil, make a 3-inch/7.5-centimeter tube, or "chimney," to fit in the hole and wrap a ring of your reserved dough around the chimney to support it. Return the pâté to the oven and bake until an instant-read thermometer inserted in the center reaches 140 degrees F./60 degrees C. Let cool to room temperature.

12. Pour the aspic through the vent hole, filling the terrine up to the top. Refrigerate for 12 to 24 hours.

Yield: 24 slices; 12 appetizer servings

[NOTES: 1. If you're cutting the salmon yourself, you'll have a lot of trimmings. Use them for or add them to a seafood sausage (page 146) or salmon pâté (page 240), or cure and hot-smoke (see pages 52 and 74) for a smoked salmon spread, mixing to taste with some butter, crème fraîche, sour cream, lemon juice, and fresh herbs. 2. See pages 203–210 for a detailed description of the general terrine method.]

Aspic for Seafood Terrines

THE STOCK

3 pounds/1.5 kilograms bones from lean white-fleshed fish
      (Dover sole, turbot, snapper, grouper)
1 tablespoon/15 milliliters vegetable oil
½ cup/70 grams chopped celery
½ cup/70 grams chopped onion
½ cup/70 grams chopped leek (white part only)
1 bunch fresh thyme
½ bunch fresh flat-leaf parsley
6 black peppercorns
¼ cup/60 milliliters dry white wine
5 cups/1.25 liters cold water

THE ASPIC

6 tablespoons/90 milliliters cold water
1 ounce/36 grams powdered gelatin (about 3 tablespoons)
1 quart/1 liter fish stock (above), chilled
2 large egg whites
1 ¾ cups/250 grams finely chopped ripe tomatoes
1 tablespoon/6 grams roughly chopped fresh parsley
4 ounces/110 grams shrimp, shelled, deveined, and finely
      chopped
Juice of ½ lemon

1. *To make the stock:* In a nonreactive pot, sauté the fish bones in the oil over medium heat. Add the vegetables and sauté for 8 to 10 minutes, until softened. Add the herbs, peppercorns, wine, and water and bring to a simmer, then reduce the heat to a bare simmer for 45 minutes. Do not allow the stock to boil, or it will become cloudy and be difficult to clarify.

2. Strain the stock through a fine-mesh sieve or a strainer lined with cheesecloth. Cool, then refrigerate until chilled.

3. *To make the aspic:* Place the cold water in a bowl, sprinkle the gelatin over, and allow it to absorb the water (bloom). Set aside.

4. In a large heavy-bottomed nonreactive saucepan, add the cold stock and the remaining ingredients, including the bloomed gelatin. Raise the heat to medium-high, and bring to a simmer, stirring continuously with a wooden spoon so the egg white doesn't stick to the bottom, until a raft forms. Once the raft forms, lower the heat, stop stirring and gently simmer for 30 minutes; don't allow to boil, or the stock will be cloudy.

5. Strain the stock through a dampened cheesecloth or a coffee filter. Put a few tablespoons on a plate and chill it to test the strength. The aspic should be firm enough to cut but melt when it hits the warmth of your mouth. If it is too loose, bloom more gelatin, dissolve it in the warmed stock, and test for strength again.

Yield: 3 cups/750 milliliters

GRILLED VEGETABLE TERRINE WITH GOAT CHEESE

This is a salad–vegetables with a vinaigrette–in terrine form. The only difference is that here the vegetables are molded and the sauce is mixed with gelatin to bind the terrine and hold its shape. Vegetable terrines should be intensely flavored and visually dramatic; because they're also healthful and refreshing to eat, they're invariably popular. This one can be served with additional vinaigrette (you might add some roasted shallot for a roasted shallot vinaigrette). If you're serving it as an elegant first course, you might rest it on a thin circle of Basil Cream Sauce (page 293). But it's fantastic all by itself.

This recipe uses late-summer vegetables, those used in a ratatouille, but you can include others if you wish. Shiitake mushrooms would work, as would portobellos; roasted red peppers would be dramatic; a layer of carrot would be a nice visual element as well. Green beans would add an interesting circular element to each slice, soft herbs a bright fresh flavor. You might blanch some chard or spinach or leeks to line the mold and make the exterior of the terrine more visually appealing especially if you chose to broil rather than grill the eggplant; distinct grill marks on the eggplant add to a handsome appearance.

Whatever vegetables you use, be sure to cook them until they are tender and delicious. If they're not delicious to eat by themselves, they're not going to get better in the terrine.

Use a small traditional terrine mold for this recipe, not much wider than 3 inches/7.5 centimeters. A larger mold would result in an unwieldy terrine that's difficult to slice neatly.

1 eggplant (1.5 pounds/675 grams), peeled and sliced lengthwise into ⅛-inch/0.25-centimeter slices

2 zucchini (about 1 pound/450 grams), sliced lengthwise into ⅛-inch/0.25-centimeter slices

2 yellow squash (about 1 pound/450 grams) sliced lengthwise into ⅛-inch/0.25-centimeter slices

½ cup/125 milliliters olive oil

Kosher salt and freshly ground black pepper to taste

3 tablespoons/45 milliliters water

2 teaspoons/8 grams powdered gelatin

½ cup/125 milliliters balsamic vinaigrette (see page 287) or vinaigrette of your choice

2 green bell peppers, roasted, peeled, seeded, and cut into thin strips (see page 129)

8 ounces/225 grams goat cheese, softened

3 All-Night Tomatoes (recipe follows; or substitute roasted red pepper or sun-dried tomatoes)

1. Heat the grill or preheat the broiler.

2. Toss the eggplant, zucchini, and yellow squash with the olive oil and season with salt and pepper. Grill or broil, turning once, until tender. Transfer to a rack or plate to cool.

3. Put the water in a small saucepan, sprinkle the gelatin over it, and allow to absorb the water (bloom). Then heat the bloomed gelatin over low heat until it is dissolved. Add it to the vinaigrette and keep the vinaigrette in a warm place.

4. Line a terrine mold (12 by 3 by 3 inches/30 by 7.5 by 7.5 centimeters) with plastic wrap, leaving enough overhang on the two long sides to cover the filled terrine (moistening the mold first will help the plastic wrap adhere to the corners). Lay the eggplant slices crosswise in the mold so that the ends hang over the sides (they'll be folded over to seal the terrine at the end). Brush them with vinaigrette. Repeat the process with the zucchini, then the yellow squash, allowing them to extend over the sides. Lay the strips of green pepper in the mold and brush with more vinaigrette. Gently press the softened goat cheese evenly into the mold. Lay the tomatoes on the goat cheese and brush with vinaigrette. Fold the eggplant-squash flaps over the top and brush with any remaining vinaigrette. Fold the plastic wrap over the top to seal and refrigerate overnight.

5. Remove the terrine from the refrigerator about half an hour before serving. To serve, open the top flaps of plastic and turn the terrine out onto a cutting board. Cut into ⅜-inch/0.75-centimeter slices.

Yield: 24 slices; 12 appetizer servings

All-Night Tomatoes

Tomatoes are composed mainly of water. Dehydrating tomatoes in a low oven greatly intensifies their flavor–tomato to the power of ten–yet they remain relatively tender, not dense and chewy like sun-dried tomatoes. Brian uses these tomatoes in the vegetable terrine, but they'd also be great in a salad, added to pasta, or as a garnish for chicken or fish. The recipe can be easily multiplied.

3 small Roma (plum) tomatoes
2 tablespoons/30 milliliters extra virgin olive oil
Kosher salt and freshly ground black pepper as needed

1. Cut the tomatoes lengthwise in half and remove the cores. Place them in a bowl and toss with the olive oil and salt and pepper to taste.

2. Place the tomatoes cut side down on a rack set on a small baking sheet. Place the tomatoes in the oven set and turn it on to 90 degrees F./32 degrees C. If you're unable to set the temperature that low, set it on the lowest setting and leave the oven door slightly ajar. Leave the tomatoes in the oven overnight to dehydrate.

3. Store the tomatoes covered in the refrigerator for up to a week.

Yield: 6 tomato halves

AVOCADO AND ARTICHOKE TERRINE
WITH POACHED CHICKEN

This summer terrine is colorful, flavorful, light. It combines two vegetable mousses with poached chicken breast, but it can also include chunks of sautéed foie gras, or even chicken liver, if you wish. There are several steps to making the components that go into this terrine, so allow plenty of time to make it. The terrine is excellent with a mâche salad seasoned only with extra virgin olive oil and salt (vinegar would distract from the terrine's delicate flavors).

THE GARNISH

One 5-ounce/140-gram boneless, skinless chicken breast

1 large tomato (about 8 ounces/200 grams), peeled

1 tablespoon/18 grams minced shallot

1 teaspoon/6 grams minced garlic

1 tablespoon/15 milliliters olive oil

Kosher salt and freshly ground black pepper

1 pound/450 grams fresh foie gras (optional), cleaned

8 baby spinach leaves

THE ARTICHOKE MOUSSE

3 tablespoons/36 grams powdered gelatin

9 tablespoons/135 milliliters cold water

1 ½ cups/375 milliliters heavy cream

6 large artichoke bottoms, cooked (see page 249), patted dry

THE AVOCADO MOUSSE

2 tablespoons/24 grams powdered gelatin

6 tablespoons/90 milliliters cold water

1 cup/250 milliliters heavy cream

5 avocados, peeled and pitted

1. Poach the chicken in gently simmering stock (see page 226) or water to cover to an internal temperature of 160 degrees F./71 degrees C., being careful not to overcook it. Let it cool, then cut horizontally in half to make two thin cutlets about 3 inches/7.5 centimeters by 4 inches/10 centimeters by ½ inch/1 centimeter. Refrigerate until ready to use.

2. Quarter the tomato and slice out the seed and inner flesh so that you are left with tomato "petals."

3. Sauté the shallot and garlic in the olive oil in a small sauté pan over medium heat until softened but not browned. Add the tomatoes and cook for 1 to 2 minutes. Season with salt and pepper, then transfer to a plate to cool.

4. If using foie gras, preheat the oven to 300 degrees F./150 degrees C. Heat a sauté pan over medium heat. Season the foie gras, and brown on all sides. Transfer the foie gras to the oven and roast to an internal temperature of 135 degrees F./55 degrees C. Remove the foie from the pan (reserve the rendered fat for another use), drain on paper towels and allow to cool.

5. *To make the artichoke mousse:* In a small bowl, sprinkle the gelatin over the water and allow it to absorb the water (bloom).

6. Heat the cream in a medium saucepan over medium heat until warm. Add the bloomed gelatin and the artichoke bottoms and bring to a boil, stirring to dissolve the gelatin. Transfer to a blender or food processor and puree until smooth. Pass the mixture through a drum sieve, or tamis, into a bowl. Cover to keep warm, so the gelatin won't set, and set aside.

7. *To make the avocado mousse:* Bloom the gelatin as above. Heat the cream in a medium saucepan. Add the avocados and bloomed gelatin and bring to a boil, stirring to dissolve the gelatin.

8. Transfer to a food processor and puree. Pass the mixture through a drum sieve, or tamis, into a bowl. Cover to keep warm so the gelatin won't set.

9. *To assemble the terrine:* Line a 1½-quart/1.5-liter terrine mold with plastic wrap, making sure there are as few wrinkles as possible and leaving enough overhang on the two long sides to cover the filled terrine (moistening the mold first will help the plastic adhere). Using a spatula, spread half the artichoke mousse evenly in the mold. Lay 4 spinach leaves across the top and press down gently so the artichoke mousse starts to push over the top of the leaves, then use the spatula to smooth the mousse over the spinach. Lay half the tomatoes down the length of the terrine. Spread half the avocado mousse evenly over the tomatoes. If using the foie gras, place on top and press down so the mousse comes up around it, then smooth with the spatula.

10. Place the chicken breasts end to end in the terrine, and cover with the remaining avocado mousse. Lay the remaining spinach leaves and then the tomatoes on top, smoothing the mousse as above, then cover with the remaining artichoke mousse.

11. Fold the overhanging plastic wrap over the top of the terrine to cover, and refrigerate overnight.

Yield: 24 slices; 12 appetizer servings

TO PREPARE ARTICHOKE BOTTOMS (HEARTS)

6 artichokes
1 pound/450 grams onions thinly sliced
2 lemons
¼ cup/60 milliliters white wine vinegar
2 bay leaves

1. Cut off the top third of each artichoke. Peel the stems.

2. Combine all the ingredients in a pot just large enough to hold the artichokes comfortably. Add enough water to cover and bring to a boil over high heat. Reduce the heat to a simmer, and place a heatproof plate or a lid perforated pan on the artichokes to keep them submerged. Simmer until tender, about 1 hour. Remove the artichokes from the cooking liquid and drain well (discard the cooking liquid).

3. Remove all the leaves from each artichoke. With a spoon, scoop out the choke (the hairy inner part) and discard. The artichoke bottoms will keep for 2 days, covered, in the refrigerator.

Yield: 6 artichoke bottoms

HEADCHEESE

Headcheese is not cheese. This delicacy comprises pieces of cooked meat from a calf's or a pig's head combined with the gelatinous cooking liquid. Once cooled and formed in a mold it is easily sliced and eaten at room temperature. In England this is referred to as brawn and in France as *fromage de tête.*

For this traditional headcheese, succulent meats are cooked till tender, then cooled and packed with fresh herbs into a terrine mold, covered with the gelatin-rich stock used to cook the meats, and chilled. Brian and I treat the meats with a basic brine and pink salt for the cured flavor and the bright pink color the salt gives the meats; because it's added for color and flavor only, not for safety, it may be omitted. Also note that the sweet spices, nutmeg and allspice, are optional; if you prefer a more straightforward, less spicy flavor, omit them.

The headcheese can be sliced and eaten as is, with good crusty bread and some mustard. It can also be sliced and served as a first course with greens and a vinaigrette as a salad.

1 pig's head (available at specialty butchers or by mail-order: see Sources, page 304)

4 fresh pig's trotters or hocks (about 6 pounds/3 kilograms total)

2 gallons/8 liters All-Purpose Brine (page 59), made with the addition of 4 ounces/112 grams (10 teaspoons) pink salt (optional), chilled

1 cured pork tongue (about 1 pound/450 grams)

2 cups/500 milliliters dry white wine

1 bouquet garni (leek, bayleaf, fresh thyme, and parsley tied together with butcher's string)

4 garlic cloves

10 black peppercorns

8 bay leaves

6 cloves

Kosher salt if needed

Freshly grated nutmeg (optional)

Ground allspice (optional)

1 tablespoon/15 milliliters white wine vinegar

1. Place the pig's head and trotters in the chilled brine, weighting them down with a plate to keep them submerged, and refrigerate overnight.

2. Remove the pig's head and trotters and rinse well; discard the brine. Place the head and trotters, along with the tongue, wine, bouquet garni, garlic, and spices in a large deep stockpot and cover with water by about 2 inches/5.1 centimeters. Bring to a simmer, then reduce the heat and simmer very gently (between 180 and 190 degrees F./82 to 87 degrees C.), skimming occasionally, for 3 hours, or until all the meats are tender. (When the jaw bone detaches easily, the head is done.)

3. Remove all the meats from the pot and set aside to cool slightly.

4. Strain the cooking liquid through a fine-mesh strainer or a sieve lined with cheese-cloth. Skim off the fat. Spoon a few large spoonfuls onto a plate and chill to check the strength of the gel. It should be firm but not rubbery or hard. If it slides around on the plate, or if it's so soft it doesn't spring back when pressed, reduce the liquid by about one-quarter and retest. It should be a sliceable gel, but not hard as a rubberball. Taste for seasoning and add salt if necessary, as well as nutmeg, allspice, and/or a splash of vinegar, if desired.

5. Remove all the meat from the head and trotters and cut into ¾-inch/2-centimeter dice. Peel the skin off the tongue; discard the skin and cut the meat into ½-inch/1-centimeter dice.

6. Line a 1½-quart/1.5-liter terrine mold with plastic wrap, leaving enough overhang on the two long sides to cover the filled terrine. Combine all the meat in the mold, and pour enough of the cooking liquid over to just cover. Fold the plastic wrap over the top and press down to make sure all the ingredients are covered. Refrigerate overnight, or for up to a week.

Yield: 24 slices; 12 appetizer servings

7.

━━◦━▷●◦ ◦●◁━◦━━

FAT:

THE PERFECT COOKING ENVIRONMENT

. . . .

Slow-cooking poultry, pork, or even fish submerged in fat, a technique generally referred to as confit, may be the best possible way to cook it. Fat is dense and flavorful, the perfect cooking medium for a leg of duck or a chunk of pork belly. Fat that's solid at room temperature also becomes the perfect environment in which to store your cooked meat, protecting it from oxygen and light. And, finally, the fat ensures that when the meat is reheated, it remains moist and succulent. The following recipes illustrate the basic confit technique, followed by numerous variations on the theme.

The literal translation of the word *confit* is "preserved." When the word is applied to a type of meat, it means poached in fat and, strictly speaking, stored within that fat until it's ready to use. Because it was originally a preservative technique, confit falls within the charcuterie rubric, and it is perhaps the most accessible and easy charcuterie technique for the home cook. It is also the most delicious and the most versatile. Once again, we no longer need this method to save our food from going bad, but we use it because it tastes exquisite.

In this country, confit is almost always associated with duck or goose, but most any meat can be confited, and many chefs use the term to apply to any meat or fish poached in fat. Ducks and geese may have been the raison d'être of the confit. When it came time to harvest the foie gras, the valuable fattened liver from specially raised birds, French farmers would have had far more meat than they could eat or sell. Happily, the birds produced extraordinary amounts of fat. So they first cured the meat with salt, then poached it in its own fat and left it to cool submerged in that fat. Prepared this way, and stored in the pot in which it was cooked, the duck (or goose) would last for months in a cool cellar, provided the fat stayed solid to prevent air from reaching the duck. Quantities of meat could thus be preserved and eaten throughout the year. The fat would also be used repeatedly, first to cook the duck and then to fry potatoes.

Pork is excellent for confit. In fact, confiting a pork loin from a commercially raised hog is a superb way of cooking this debased cut of meat. Actually, any part of the pig can be confited with excellent results. Confited belly and shoulder are rare treats in the home kitchen, extraordinary for their succulence and flavor.

Moreover, confit is extraordinarily versatile. Duck confit may be best with the skin simply crisped up in the oven and eaten as is, but it can also become a part of a cassoulet, a traditional French bean stew. Its rich, spicy succulence is an exquisite contrast to greens and a vinaigrette. Shredded, it can become ravioli filling or be rolled up in a crepe, added to a soup, or pounded into rillettes.

As the duck poaches in the fat, the seasonings flavor the fat, transforming it into an extraordinary cooking medium, whether for more confit, for sautéing, or even for seasoning. The fat can be used in a warm vinaigrette, drizzled over bitter greens, or added as an enriching touch to sauces.

The process also results in an amazing essence, "confit jelly." When you confit meat and bones, they release juices and flavor and collagen–just as they do when simmered in water to make stock. The juices fall to the bottom of the pot they've been cooked in, under the fat. But rather than being diluted, as in stock, they become concentrated. When this cold rubbery substance is removed, you have powerfully flavored, salty, gelatinized broth. Simply melted and served with the duck, it's delicious. It can be added to a vinaigrette to flavor it, used to fortify a sauce, or added to rillettes made from the confit.

This chapter offers both traditional confit recipes along with more unusual confits, such as those for pork loin, shoulder, and belly. Recipes for rillettes are included as well, as they are a charcuterie technique and a cousin of confit. In classic rillettes, pork, duck, goose, or rabbit is poached in stock or fat till it is falling-apart tender, then pounded with seasonings and fat into a coarse paste and spread on crusty bread.

Chefs often use the term *confit* with vegetables, to describe onions, tomatoes, or fennel cooked in oil until meltingly tender. The technique is not really a confit, but it shares confit's gentle cooking and tender unctuousness. Confitures, of course, are the confits of the fruit world, in which sugar replaces the fat to make jams and other preserves.

Basic Duck Confit

Variations on duck confit are infinite, depending on the seasonings you add to your dry cure. Some people prefer the pure flavor of the duck with a little thyme, others enjoy the traditional sweet spicing of clove. The only two essentials of traditional confit are a salt dry cure–we use up to 2 percent of the weight of the meat (.3 ounces per pound/20 grams per kilo)–and gentle poaching in fat. Other than that, it's a matter of taste.

Brian and I like a little bit of cloves, black pepper, savory garlic, and some bay leaf, but even salt alone is fine. We've included variations on the seasoning, as well as a recipe for goose confit.

You have many options when it comes to the duck, and all work well: Pekin (Long Island), Muscovy, or moulard. Excellent duck legs are available online for reasonable prices (see Sources, page 305). The easiest way to make duck confit at home is to buy duck legs via mail-order (see Sources), along with a couple pounds of duck fat. This way you're assured of excellent duck and enough cooking fat to cover the duck. Many cities have markets where farmers sell duck. Order a big batch ahead of time and do all your confiting at once, so you have some all winter. Well wrapped and stored in the freezer, the fat will keep for six months or longer, and it can be reused for confit many times (eventually it will become too salty).

If you are an ambitious cook and like to work from scratch, a less expensive method is to buy a whole duck at the supermarket and remove the leg-thigh pieces from the carcass, trimming all the fat and saving it for rendering. We've never had a problem with the amount of fat rendered from a duck, but if you find that your fat won't cover the duck add shortening or, better, lard or olive oil to cover. You can confit the breasts as well. This method also leaves you with a carcass to roast for making stock. So buying a whole duck gives you quite a bit more for the money, if you like to work in the kitchen.

TRADITIONAL COOKING

To cook confit, preheat your oven to its lowest setting, ideally 180 degrees F./82 degrees C. and no higher than 200 degrees F./93 degrees C. Confit can become tough and stringy if the oven is too hot. Rinse the salt and seasonings from the cured duck and thoroughly dry with paper towels. Submerge the legs in rendered fat in a pot just big enough to hold the legs snugly, bring the fat just to a simmer on the stovetop, and then place the pot in the low oven for 6 to 10 hours. The confit is done when the legs, resting on the bottom of the pan, are very tender. The fat should be clear, not cloudy; this clarity indicates that the duck is no longer releasing juices, and that the juices that have been released are cooked and resting on the bottom of the pot. Remove the pot from the oven, and allow the duck to cool to room temperature.

The fat, separated from the confit jelly (see page 255), can be reused several times before it becomes too salty. The fat will, of course also pick up the seasonings used, so depending on the seasonings you're using the next time, you may need fresh fat.

CONFIT USING SOUS VIDE

Cooking food sous vide, in a vacuum sealed bag, has become increasingly popular. It is indeed an excellent method of creating the confit effect using considerably less fat. It can be used for any meat you wish to confit, from pork belly to duck legs to chicken thighs.

The method is simple: rub the meat with cure and seasonings of your choice (again we recommend 1.75 to 2 percent salt by weight), add about a quarter cup/60 milliliters of fat, seal it all in the bag, and refrigerate it overnight. Preheat a waterbath to 180 degrees F./82 degrees C. Drop the meat into the water bath and cook it for 8 to 12 hours. Remove the bag to an ice bath and chill completely, then refrigerate it in the bag until you're ready to use it.

Certainly it's possible to chill the legs overnight in the cooking pot and reheat them to eat the next day, but letting them rest, or "ripen" as some chefs say, for a week or longer will improve their texture and flavor. To store them for a short period—a week to ten days—simply refrigerate them in the pot or transfer them to a storage container, pour the rewarmed fat over the legs, and refrigerate. We recommend transferring the duck to a storage container, preferably an earthenware crock that will keep light out. Ladle the hot fat over the legs until they are completely submerged, being careful not to pick up any juices that have settled in the bottom of the pot (the jelly), which can sour over time. Once the fat has congealed, place a layer of plastic wrap directly on top of the fat, so that no air reaches it, cover with a lid or with foil, and store in the back of your refrigerator for up to 6 months.

To serve the confit, remove it from the refrigerator several hours before reheating it to allow the fat to soften; the duck is delicate and can tear if you try to remove it from fat that's too firm. To reheat the duck, arrange the pieces skin side down in an ovenproof sauté pan and heat over medium heat until the skin begins to render its fat and get crispy, about 5 minutes. Then turn them skin side up and finish them in a hot oven (425 degrees F./220 degrees C.), 5 to 10 minutes. Or if the skin is not too thick, you can simply reheat in the hot oven for 10 to 15 minutes. For the best confit, the skin must become crisp. You can also finish them skin-side-up beneath a broiler to ensure crisp skin.

LE VRAI CONFIT: TRADITIONAL CONFIT DE CANARD

On a recent trip to Gascony in southwestern France, we were re-inspired by *confit de canard*, which is a way of life and an emblem of the food culture there. In Gascony, households do not confit duck to eat immediately or later that week with a salad, but rather to save for consumption throughout the winter. In some cases confit—pig, duck, and goose—was so critical to a farmer's winter food supply, the family would put confit up for two or even three years, eating the two-year-old confit only if the one-year confit and the current year's supply were good, according to Kate Hill, an American who has lived in Gascony for twenty-five years.

Kate runs a culinary retreat in her eighteenth-century stone farmhouse near the city of

Agen, and offers a monthlong charcuterie program for people who are serious about the art and craft of this specialty (we highly recommend her program, by the way).

It was while discussing duck confit over a duck confit lunch that I determined to include, with Kate's help, a true French confit preparation, one designed specifically for purposes of preservation. It must include no garlic or herbs or seasonings other than salt, to eliminate any chance of contamination from exterior sources (garlic, for instance, can harbor botulism spores, which aren't killed in the cooking and could then result in generating its toxin in the anaerobic environment of the fat).

Kate explained that, traditionally, a whole duck would be cut into pieces and every bit of it would go into the confit pot, not just the legs. The big breast, wings, and gizzard would be added later in the cooking, and the carcass would go in at the end, so that the remaining meat could be picked off to make rillettes, seasoned with the confit jelly, the gelatin by-product of the cooking process.

Because most people didn't have temperature-controlled ovens, confit was cooked on the stovetop, gently, frequently stirred and carefully looked after.

Before canning, the method was to cook the duck, lift it out of the fat, and put it into the confit crock, ladle pure fat over it to cover, allow the fat to solidify, then put a layer of harder pork fat on top of this, and finally to cover it with some opaque paper. It would be good for the year and more; it was preserved, which is what *confit* means.

Canning (or jarring) confit is common now, and a great way to keep your duck indefinitely if you like to preserve this way. It's best to cut off the ends of the drumsticks after they've cooked to give you more room in the vessels.

However you want to preserve your duck, we urge those who want to put duck up for a year or more to follow these instructions and taste what truly ripe confit is all about. You can use this method with whatever duck is available to you but if you're going to do it properly, you should only use the breed of duck common in Gascony, the Moulard, which grows very big and has plentiful fat. Their breasts, called *magret*, are richer and more flavorful than strip steaks if you grill them. It's the breed that is used to create foie gras (which Kate is quick to point out was originally not a specialty item, but rather a happy by-product of a fattened duck—it was the fat, not the foie, farms were after).

Moulard duck legs and *magret* are available from a Gascon ex-pat, Ariane Daguin, who created a company called D'Artagnan and sells numerous duck, foie gras and other Gascon foods from her base in New Jersey (see Sources, page 305). We highly recommend her products, and we also recommend those of Hudson Valley Foie Gras, which also sells Moulard duck and duck fat (they and Daguin work closely together).

One final note: If you're nervous about bacterial issues or don't feel comfortable with room temperature preservation, follow these instructions and store at refrigerated temperatures. If you remove the legs from the fat and you see green or black mold, you've let an air pocket remain trapped, so evaluate using your own common sense; bad technique or inadequate sanitation can ruin a whole batch if you're not careful. The legs should look like they were just cooked when they come out of the crock, with a clean coating of snowy-white fat.

Moulard duck, as you have (legs, whole duck cut into pieces, gizzards)
Coarse sea salt or gray salt
Extra duck fat as needed if not using whole ducks
Pork fat (optional)

1. Put your duck in the vessel it will cure in. Measure a scant tablespoon of salt for each leg or leg-sized piece of confit. (If you'd like to be precise, use 3 percent salt relative to the weight of the meat.) Rub the salt into the meat well so that it starts to dissolve. Arrange the duck fat side down so that the meat side cradles the released juices and cover with plastic or a kitchen towel. Allow it to sit overnight, either refrigerated or in a cool or cold place (this is a matter of where you are and how cold it is; in Gascony the fall nights are cool; if you live in a hot climate, it's a good idea to refrigerate your duck). The duck should cure for 12 to 16 hours or so.

2. *If you're confiting a whole Moulard*, you will have enough fat for the confit when you break the duck down. Take all the skin and fat not naturally connected to the pieces you're confiting, chop it up, and put it in a sauce pan with a quarter cup of water over medium-high heat. When the water boils, turn the heat to low and continue to cook until all the fat has rendered and all you have left are delicious duck cracklings, which can be salted and eaten warm or used as a garnish on salads or potatoes.

3. *If you are confiting duck legs*, you'll also need to buy duck fat, about 6 cups/1.5 liters for 8 legs, depending on your cooking vessel.

4. Put the duck into your pot, cover it with fat, and cook it gently, about 180–190 degrees F./82–87 degrees C., until it's tender, with a few bubbles coming to the surface. Kate explains that the time cooking depends on how you are preserving it. If you are preserving in jars, the total cook time should be about 1½ hours, because it will cook for longer in the jarring process. If you are preserving it in an unsealed vessel, then you should cook your duck for about 3 hours. Duck can go as long as 10 hours in a low oven at 180 degrees F./82 degrees C., which is common in America; this ensures that the confit will be meltingly tender and suc-

culent. But in pre-refrigeration days the cook wouldn't have spent 10 hours and that much fuel to cook duck. So 2 to 3 hours is traditional. All of the *confit de canard* we had in Gascony had a solid bite to it, rather than being falling off the bone tender. This is how it's done there, consistent with the rugged Gascon spirit.

5. *To store it in an earthenware vessel*, place the cooked duck in an immaculately clean vessel; some people use some sort of rack or saucer in the bottom of the vessel to keep the duck from touching the bottom and thus ensuring that as much of the duck surface as possible is against fat. Strain the fat through a fine mesh sieve into a separate bowl. Allow the gelatin to settle. Then ladle pure fat over the duck to cover. Allow the duck fat to cool and solidify, either in a cold cellar or refrigerator. For an extra measure of protection you may want to put a layer of pork fat over the duck fat. Allow it to solidify, cover the vessel with opaque paper, and tie a string tightly around the outside of the vessel to hold the paper tightly on the vessel.

6. *If you're jarring the confit*, place the duck in sterilized jars, seal them, and process them submerged in boiling water for 2 hours to further sterilize and create an airtight seal. Remove them from the heat. When cool enough to handle–they can sit overnight if need be; they're completely pasteurized–store at room temperature in the coolest, darkest part of your dwelling indefinitely.

7. This results in the traditional Gascon *confit de canard*. It can be used in a cassoulet or bean stew, or can be sautéed, roasted, or broiled as it is in the summer and served with potatoes cooked in duck fat. Kate notes that chef André Daguin, Ariane's father, recommends steaming the confit in a couscousiere to tenderize it and ready the skin to become extra crisp. For those at home who have the refrigerated storage space, we recommend you store your confit here for at least 6 months and preferably a year to develop the silky texture and aged succulence of traditional *confit de canard*.

DUCK CONFIT WITH CLOVE

You can serve duck confit in many ways. It's wonderful on a salad with a sharp vinaigrette (some of the salty confit jelly blended in here is excellent). It goes well with green vegetables such as spinach or Brussels sprouts. It's wonderful in risotto, superb in soups and stews, and, shredded, it can be a filling for pasta or a corn tortilla, or it might be used as garnish in a cold pâté. Of course it's perfect rested atop diced potatoes fried in duck fat.

6 Pekin (Long Island) duck legs, about 5 pounds/
    2.25 kilograms
3 tablespoons/40 grams kosher salt
4 whole cloves
6 black peppercorns
3 garlic cloves, sliced
3 bay leaves
2 to 4 cups/500 to 1000 milliliters rendered duck fat or lard
    (see page 262) or a combination

1. Sprinkle the duck legs all over with the salt, and put them in nonreactive container.

2. Roughly crush the cloves and peppercorns with the side of a knife and scatter evenly over the duck. Press some sliced garlic onto each duck leg. Break the bay leaves in half and press a half-leaf onto each piece of duck. Cover with plastic wrap and refrigerate overnight, or for up to 48 hours.

3. Rinse the duck under cold water, wiping off all the seasonings. Pat dry.

4. Preheat the oven to 180 to 200 degrees F./82 to 93 degrees C. (depending on how low you can set it).

5. Place the duck legs in a pot; a 6-quart stockpot or a Dutch oven works well for this. The legs can be in one layer or two layers; the only critical factor is that you have enough fat to completely cover the duck. Cover the legs with fat and bring to a simmer over medium-high heat. Then place, uncovered, in the oven and cook for 6 hours, or until the legs are completely tender and have settled on the bottom of the pot and the fat has become clear, up to 10 hours.

6. Remove from the oven and cool to room temperature in the pot, then refrigerate; be sure the duck is completely submerged in the fat. Store covered in the refrigerator for up to a month.

7. To serve the confit, remove it from the fridge several hours ahead to allow the fat to soften. Preheat the oven to 425 degrees F./220 degrees C.

8. Remove the legs from the fat. Place them in a baking pan or on a baking sheet, and roast until the meat is warmed through and the skin is crispy, 15 to 20 minutes. Or gently sauté them to crisp the skin, then finish in the oven.

Yield: 6 servings

1 pound duck or goose skin and fat, scraps of pork fat, or suet, roughly chopped

1. Combine the fat and ¼ cup/60 milliliters water in a heavy-bottomed saucepan and set it over very low heat, uncovered, for several hours to render. The fat will liquefy and the water will evaporate, leaving pure fat. Don't let the fat come to a boil or turn brown, or it will taste harsh.

2. Strain the fat through a sieve lined with cheesecloth. Let cool, then store covered in the refrigerator. The golden brown skin–cracklings–remaining in the cheesecloth can be saved to use as a flavoring for braised cabbage or sautéed potatoes.

Yield: 1 cup/250 milliliters

DUCK CONFIT WITH STAR ANISE AND GINGER

This cure will give the duck a distinctive seasoning and demonstrate the range of flavors that can be brought to a piece of duck. However, the seasoning of orange, anise, ginger, and scallions is subtle; confit should not bring potpourri to mind when you eat it.

6 Pekin (Long Island) duck legs, about 5 pounds/
2.25 kilograms
3 tablespoons/40 grams kosher salt
2 large star anise, ground to a powder in a spice grinder
¼ teaspoon/1 gram ground cinnamon
3 scallions, white and pale green parts only, roughly chopped
3 garlic cloves, roughly chopped
1 tablespoon/10 grams minced fresh ginger

Grated zest of 1 orange

4 black peppercorns, cracked beneath a small heavy pan or
side of a knife

2 to 4 cups/500 to 1000 milliliters rendered duck fat or lard
(see page 262) or a combination

1. Sprinkle the duck all over with the salt. Distribute the remaining ingredients evenly on the duck, cover, refrigerate overnight; or for up to 48 hours.

2. Rinse the duck well, wiping off all the seasonings, and proceed as directed in the recipe for Duck Confit with Clove (page 260), starting with step 4.

Yield: 6 servings

GOOSE CONFIT

This is a peppery-savory cure that goes especially well with goose, though you could certainly use it for duck too. There's quite a bit more meat than on a duck, which is why goose is commonly prepared as confit. The abundant meat also makes goose a good choice for rillettes (to transform a confit into rillettes, see page 267). The fat, which will take on the seasonings of the goose, can be reused several more times, or until it becomes too salty; it can be frozen for up to 4 months, it is also fantastic for frying potatoes.

Don't hesitate to ask the butcher to prepare the goose for you. Order a fresh goose and ask that it be cut into bone-in pieces: two leg-thighs, two breasts, and two wings, with wing tips removed and reserved.

1 goose, about 12 pounds/5.5 kilograms

7 tablespoons/90 grams kosher salt

4 bay leaves

8 garlic cloves, crushed

1 bunch fresh thyme

6 allspice berries

3 tablespoons/30 grams black peppercorns

4 whole cloves

3 quarts/3 liters rendered goose fat or lard or a combination
(see page 262)

1. Cut the goose into 6 pieces, 2 breast halves, 2 leg-thighs and two wings, trim the loose fat and reserve all the fat.

2. Combine all the remaining ingredients except the goose fat or lard in a spice mill and pulverize. Rub the mixture over the goose pieces, place them in a nonreactive pan, and refrigerate, covered, for 2 days.

3. Preheat the oven to 180 to 200 degrees F./82 to 93 degrees C.

4. Rinse the goose pieces under cold water, wiping off the seasonings, and dry thoroughly with paper towels. Submerge the pieces in the rendered fat in a stock pot or Dutch oven; all the meat must be completely covered. Bring the fat to a gentle simmer on top of the stove, then place in the oven and cook until the meat is fork tender and the fat is clear, about 6 to 10 hours.

5. Allow the goose to cool in the pot, then refrigerate, completely submerged in the fat. Or gently transfer the goose to a container in which the pieces fit snugly, and ladle the fat into it until the pieces are completely covered. Refrigerate until the fat is solid, or for up to a month; or freeze for up to 4 months.

6. When ready to serve the confit, allow the fat to soften at room temperature for several hours, then remove the goose from the fat, wiping off the excess. Sauté in a large pan on the stovetop, or roast in a 425-degree-F./220-degree-C. oven, until the skin is crisp and meat is warmed through, about 10 minutes. Transfer the fat to a plastic container or freezer bag and freeze for later use in another confit.

Yield: 6 servings

PORK CONFIT

Why confit pork? Because it's fantastic. To take an inexpensive tough cut of meat such as pork shoulder butt, and through your knowledge and skills as a cook transform it into something exquisite, well, that's what real cooking is. It's an effective technique for using any extra pork shoulder you might have.

> 2 tablespoons/30 grams kosher salt
> 3 bay leaves
> 4 garlic cloves, crushed
> ½ bunch fresh flat-leaf parsley, coarsely chopped

2 tablespoons/20 black peppercorns

1 bunch fresh sage

3 tablespoons/36 grams chopped shallots

½ teaspoon/3 grams pink salt

5 pounds/2.25 kilograms boneless pork shoulder butt,
   cut into 2-inch/5-centimeter chunks, or
   one 3-pound/1.5-kilogram boneless pork loin

2 to 4 cups/500 to 1000 milliliters rendered duck fat or lard
   (see page 262) or a combination

1. Combine all the ingredients except the pork and the fat in a spice grinder and pulverize to a powder.

2. Rub the mixture evenly all over the meat. Place it in a nonreactive container, cover, and refrigerate for 24 hours if you're using pork shoulder pieces, 48 hours if you're using pork loin.

3. Preheat the oven to 180 to 200 degrees F./82 to 93 degrees C.

4. Rinse the pork under tepid water, wiping off all the seasonings, and dry thoroughly with paper towels. Submerge the meat in the rendered fat in a stockpot or Dutch oven; the meat must be completely covered in fat. Bring the fat to a gentle simmer on the stovetop, then place the pot, uncovered, in the oven, and cook until fork-tender for 4 to 6 hours for shoulder, 3 hours for loin.

5. Cool in the fat, then cover, making sure all the meat is submerged in the fat and refrigerate for at least 24 hours, or for up to 3 weeks. Or freeze for up to 4 months.

6. To serve, allow the pork to come to room temperature, remove from the fat, and sauté over medium heat or roast at 425 degrees F./220 degrees C. until hot.

Yield: 4 pounds/1.75 kilograms pork confit

JIM DROHMAN'S PORK BELLY CONFIT

Jim Drohman is chef-owner of Le Pichet in Seattle, a bistro that specializes in charcuterie. Local cooks have made Drohman's pork belly a kind of a cult item within their circle, repairing after work to the restaurant for this fantastic example of confit, a variant of *rillons*, pieces of pork belly sautéed until crispy.

Jim developed his cure based on Loire Valley tradition, which introduces white wine to the cure. Other areas, he says, might use another alcohol, such as Cognac. The seasoning is a sweet-spice mix that can be used with just about any confit. The amount of cure below is enough for 6 pounds/2.75 kilograms of pork belly, quite a bit more than most people will want to prepare at home, but stored in a tightly sealed container, it will keep for months.

You could roast or sauté the pieces to reheat, but deep-frying them, as he does, is the easiest way to ensure a uniform crust and a melting texture inside (from a calorie or fat standpoint, it would be difficult to increase either through your cooking method, even if you wanted to). Also, it's usually juicier this way, as the density of the fat and the confit are similar; use either canola oil or the tasty fat you've stored the belly in for the cooking.

Jim serves this with mustard and, usually, some sort of spiced fruit preserves. In spring, perhaps feeling guilty about its high fat content, he sets it on some green beans dressed with a vinaigrette and sliced almonds. But it's still deep-fried fat no matter how you try to healthy it up.

At home, good mustard and a crusty baguette are the perfect accompaniments. You might also serve it alongside a salad with a vinaigrette.

THE DRY CURE
2 tablespoons/20 grams freshly ground black pepper
½ tablespoon/6 grams ground cinnamon
½ teaspoon/2 grams ground cloves
¼ teaspoon/1 gram ground allspice
3 bay leaves, crumbled
10 sprigs fresh thyme
4 tablespoons/50 grams kosher salt
1 teaspoon/7 grams pink salt

6 pounds/2.75 kilograms pork belly, skin removed and cut into 1-inch/2.5-centimeter by 3-inch/7.5-centimeter chunks
Dry white wine as needed
Rendered pork or duck fat as needed (see page 262)
Canola oil or rendered pork or duck fat for deep-frying

1. Combine all the cure ingredients in a bowl and stir to distribute the seasonings evenly.

2. Toss the pork with the cure to coat evenly. Pack into a nonreactive container and cover with white wine. Cover and refrigerate for 24 to 36 hours.

3. Preheat the oven to 250 degrees F./120 degrees C.

4. Remove the pork from the cure and pat the pieces dry with paper towels. Place the pork in an ovenproof pot or Dutch oven and cover with the rendered fat. Bring to a simmer on the stovetop, then place in the oven, uncovered, and cook until the pork is fork-tender, about 2 to 3 hours.

5. Remove the pork from the oven and cool to room temperature in the fat. (If you simply can't wait to eat this succulent bundle when it has finished its confit–we highly recommend chilling all confit, which intensifies the juicy tenderness of the meat–you can pour off and reserve the fat, then return the pan to the stovetop over high heat until the meat is nicely browned.) Refrigerate the pork in the pan it was cooked in or transfer to another container and add the fat; the pork should be completely submerged in fat. Refrigerate until completely chilled, or for up to 2 months.

6. To serve, remove the pork from the refrigerator, preferably a few hours ahead. Remove the pork from the fat, and wipe off the excess. In a deep heavy pot, heat the oil for deep-frying to 350 to 375 degrees F./175 to 190 degrees C. Deep-fry the pork belly until crispy and heated through, about 2 minutes if it was at room temperature. Remove and drain on paper towels.

Yield: About 12 servings

Rillettes

Rillettes are the perfect party hors d'oeuvre, an amazing midweek snack, and an example of a beautiful utilitarian preparation–flavorful, economical, satisfying, great to make ahead–an excellent technique that was once all but unheard of in the American home kitchen. Indeed, a perusal of standard home cookbooks such as *The Joy of Cooking*, Julia Child's *Mastering the Art of French Cooking*, and Craig Claiborne's *New York Times Cookbook* suggests why they might not be made often or at all at home–they're not even mentioned.

They ought to be.

Rillettes, considered a "potted" food because they are customarily served in a pot or jar or a ramekin, are some type of meat (most often pork, rabbit, or duck) that's long-simmered with herbs and aromatics and fat until it's meltingly tender. Then it's pounded or mixed to a spreadable paste, transferred to a ramekin or similar container, and sealed with a thin layer of fat. Rillettes can be refrigerated for several weeks before being set out at room temperature to be spread on toast or bread.

Techniques vary—some cooks prefer to use confited meat rather than meat simmered in stock, for instance—but the key to a rich flavor and a creamy texture remains the right proportion of fat to meat. The smooth but meaty texture is also aided by the gelatin the meat releases during the long, slow cooking.

Fish, of course, don't contain as much or the same kind of fat as meat, and so to make rillettes from fish (salmon is a common choice), the fat you add is butter, and often the crock is sealed with clarified butter.

Whether meat or fish, or even, loosely speaking, vegetable (see page 271), the other components of rillettes are simply fat and seasonings. Also characteristic of so much of charcuterie is that, packed in small elegant containers, they're the perfect portable food, whether you're bringing a dish to a party, having a picnic, or out for a day sail. Rillettes are so good we'd like to see more people serving them.

CLASSIC PORK RILLETTES

When a waiter sets down a crock of rillettes in a bistro in France, this is most likely what it will be. Pork is inexpensive and easily enhanced through cooking with aromatics, and its fat is naturally creamy. This is an excellent all-purpose version of rillettes, which is sometimes referred to as Le Mans–style, after the city in northwestern France famed for its pork rillettes.

This recipe calls for white veal stock, amazing for its neutrality and for the depth of flavor it brings to almost any meat preparation.

> 1 large leek
> 1 small bunch fresh thyme
> 3 bay leaves
> 1 celery stalk
> 8 black peppercorns
> 1 medium onion, studded with 5 cloves
> 3 pounds/1.5 kilograms boneless very fatty pork butt,
>   cut into 1-inch/2.5-centimeter dice
> Kosher salt
> 2 quarts/2 liters White Veal Stock (recipe follows) or water
> Freshly ground black pepper to taste
> About 8 ounces/225 grams rendered pork fat (see page 262)
> A small square of cheesecloth

1. Split the leek lengthwise in half, stopping about 1 inch/2.5 centimeters from the root end, and wash it thoroughly to remove any dirt or mud from between the layers. Lay the thyme and bay leaves inside the split leek, lay the celery stalk next to it, and tie them all together with butcher's string (an aromatic bundle called a bouquet garni).

2. Crack the peppercorns with the side of a knife so that they'll release their flavor, and tie them up in the cheesecloth.

3. Preheat the oven to 300 degrees F./150 degrees C.

4. Place the pork in a 6-quart/6-liter pot and cover with water to by about 2 inches/5 centimeters. Bring to a boil, then drain the pork, and rinse it under cold water (a way to quickly eliminate blood and impurities). Return the pork to the clean pot and add the bouquet garni, peppercorns, onion, 1 tablespoon/15 grams salt, and stock. Bring to a simmer, cover, and place in the oven. Cook until the meat is falling-apart tender, 4 to 6 hours.

5. Remove the pork from the pot, and set aside to cool somewhat. Strain the liquid, and set aside.

6. When the meat is cooled to slightly above room temperature, place in the bowl of a standing mixer fitted with the paddle attachment. Mix on low speed, slowly adding enough reserved liquid, until the meat shreds thoroughly and the mixture takes on a moist spreadable texture, 1 to 2 minutes. Taste and add salt and pepper if necessary (remembering that this will be served at room temperature so it should be seasoned assertively).

7. Spoon the mixture into individual ramekins or crocks. Refrigerate until chilled, then pour about ⅛ inch/0.25 centimeter rendered fat on top to seal the ramekins. Return to the refrigerator until shortly before serving, or up to 2 weeks.

8. Remove the rillettes from the refrigerator at least 2 hours before serving; they're most flavorful and easiest to serve at room temperature.

Yield: 8 portions

White Veal Stock

Veal stock is a culinary treasure that's very rarely made at home, but it's no more difficult to prepare than chicken stock. And if you've ever wondered why a great restaurant's sauces are so much more distinguished than what you make at home, veal stock may be the reason. White veal stock is neutral, meaning that it readily takes on other flavors, but it also adds substantial depth and richness to whatever dish you are preparing.

This recipe makes a good amount, but it freezes well so you have stock on hand to use at a moment's notice.

> **8 pounds/3.75 kilograms veal bones, cut into 3-inch/**
> **7.5-centimeter lengths (have the butcher do this)**
> **1 cup/140 grams finely diced onions**
> **½ cup/70 grams finely diced celery**
> **½ cup/70 grams finely diced leek—white part only**
> **2 bay leaves**
> **2 teaspoons/5 grams black peppercons**
> **1 bunch fresh thyme**
> **About 6 quarts/6 liters cold water**

1. Place the bones in a pot that is taller than it is wide, cover with water, and bring to a boil. Drain the bones and rinse under cold water. (This process cleans the bones and removes impurities that you don't want in your finished stock.)

2. Place the bones back in the pot and add the remaining ingredients except the cold water. Add the water to cover the ingredients by about 1 inch/2.5 centimeters. Bring to a boil, skimming the stock frequently, then reduce the heat to the lowest possible. Skim the stock, and simmer for 6 to 8 hours, or overnight, skimming occasionally.

3. Strain the stock through a fine-mesh sieve. Let cool, then refrigerate or freeze.

Yield: 1 gallon/4 liters

SMOKED TROUT RILLETTES

Rillettes are most commonly made from pork and pork fat, but they are also an excellent way to use smoked trout, or any smoked fish, as well as poached (nonsmoked) salmon. Rillettes made with fish use butter in place of the pork fat used when making pork rillettes (see page 267 for more on rillettes). The technique allows you to season and enhance the fish while putting it into a form that's easy to eat.

The acidity of the wine here sharpens the flavors, and lemon zest, as always, brightens them. As with most rillettes, this is best spread on toasted or crusty bread, as an hors d'oeuvre, but it also makes a good first course or, served with bread, salad, and a crisp cold white wine, a great lunch. Because rillettes travel well, they're the perfect food for picnics, tailgate parties, and other events requiring portable food.

4 ounces (1 stick)/110 grams unsalted butter

2 tablespoons/15 grams minced onion

8 ounces/225 grams boneless, skinless smoked trout fillets
    (see Note below), shredded

¼ cup/60 milliliters dry white wine

½ teaspoon/2 grams finely minced lemon zest

1 teaspoon/7 grams kosher salt

Ground white pepper to taste

2 tablespoons/16 grams finely sliced fresh chives

1. Melt the butter in a medium sauté pan over medium heat. Add the onion and sauté until soft, without browning it. Add the trout, wine, lemon zest, salt, and white pepper and cook, stirring, until well blended and most of the wine has evaporated, 8 to 10 minutes. Remove from the heat and cool to room temperature (the butter should have solidified but still be soft).

2. Transfer the mixture to a bowl, add the chives, and mix with a wooden spoon until the mixture becomes a smooth paste. Adjust the seasonings if necessary, and place in crocks or ramekins. Refrigerate until ready to serve, or for up to 10 days.

Yield: 8 portions

[NOTE: Smoked trout is available at many specialty stores. If you wish to smoke your own, brine boneless fish in the All-Purpose Brine (page 59; half the recipe will be sufficient for the amount of fish here, but add ½ teaspoon/3 grams of pink salt to the brine). Brine the fish for 4 hours, then rinse, dry, and refrigerate for 1 to 2 hours, uncovered. Hot-smoke (see page 74) to a temperature of 140 degrees F./60 degrees C., then immediately refrigerate the fish till it's thoroughly chilled.]

MEDITERRANEAN OLIVE AND VEGETABLE "RILLETTES"

Vegetarian rillettes are not, of course, traditional rillettes, which are made of meat and fat, but they are a rich, satisfying dish that, spread on crostini, could be served as a first course. Or serve as an hors d'oeuvre out of a ramekin, as you would tapenade, which is what this dish most resembles. Sweet and intensely flavored, it's a great preparation to make when these vegetables are abundant in late summer.

1 medium eggplant (1½ pounds/675 grams)

1 zucchini (8 ounces/225 grams), cut into ½-inch/1-centimeter
disks

1 yellow squash (8 ounces/225 grams), cut into ½-inch/
1-centimeter disks

4 ounces/100 grams mushrooms, quartered

2 large vine-ripened tomatoes (about 14 ounces/400 grams),
cored, quartered, and seeded

½ cup/125 milliliters extra virgin olive oil

Kosher salt and freshly ground black pepper

1 red bell pepper (6 ounces/160 grams)

1 yellow bell pepper (6 ounces/160 grams)

1 cup/140 grams diced onion

1 teaspoon/6 grams minced garlic

1 cup/200 grams Niçoise or Kalamata olives, pitted

2 tablespoons/30 milliliters balsamic vinegar

½ cup/50 grams fresh basil chiffonade (slivered)

1. Preheat the oven to 375 degrees F./190 degrees C.

2. Poke holes in the eggplant with a fork, and place on a small baking sheet. Roast until soft, about 40 minutes. Set aside to cool. Turn the oven up to 400 degrees F./200 degrees C.

3. Toss the zucchini, yellow squash, mushrooms, and tomatoes with ¼ cup/60 milliliters of the olive oil, season with salt and pepper, and spread on a baking sheet. Roast until soft and golden brown, about 20 minutes. Let cool, then finely chop and set aside.

4. Meanwhile, grill or broil the bell peppers, turning frequently, until blackened all over. Place in a bowl, cover with plastic wrap, and let cool. Then peel, seed, and dice.

5. In a small sauté pan, sauté the onion and garlic in 2 tablespoons/30 milliliters olive oil until soft but not browned; set aside.

6. Peel the skin from the roasted eggplant. Place the olives in a food processor and puree. Add the eggplant, then the onion-garlic mixture, and puree until smooth. Transfer to a medium bowl. Fold in the diced roasted peppers, zucchini, squash, mushrooms, tomatoes, balsamic vinegar, remaining olive oil, and the basil. Season with salt and pepper.

7. Store, covered, in the refrigerator until ready to use, or for up to a week. Serve at room temperature.

Yield: 1 quart/1 liter

SIMPLE RILLETTES FROM CONFIT

Once you have made confit—be it duck, goose, or pork—these rillettes can be prepared in ten minutes, and they're fantastic, probably better in fact than classical rillettes, because the meat is cooked not in stock but rather in fat. If you're going to the trouble of making confit, it's always a good idea to make more than you need so you can then have some rillettes on hand. Sealed with fat and covered with plastic wrap, then aluminum foil (fat likes the dark), they'll keep for a month or longer in the refrigerator. They also freeze well (as do most high-fat items), well-wrapped, for up to three months.

This method can be used with any meat confit and it is easily doubled. It's an excellent use for confited duck breast, which tends to be a little drier than the leg, and benefits from the added fat.

8 ounces/225 grams confit meat
¼ cup/60 milliliters confit fat, or as needed
¼ cup/60 milliliters confit jelly (see page 255), or as needed
Freshly ground black pepper
Kosher salt if needed

1. Tear the meat into chunks, discarding any skin if you're using duck, and place in the bowl of a standing mixer fitted with the paddle attachment. Add the fat and confit jelly, and season with pepper. Mix on high speed until the fat is evenly distributed and the meat is shredded and creamy looking. The rillettes should be very moist and spreadable. If the mixture seems too stiff, add more fat and/or jelly, or stock, and mix to incorporate. Taste for seasoning; the jelly is very salty, so be cautious with salt.

2. Transfer to ramekins and refrigerate. Once chilled, you can seal the ramekins with additional fat, or, because there's plenty of fat in them already, simply spread a good Dijon mustard over the top.

3. Serve at room temperature, with thin slices of toasted baguette.

Yield: Six 2-ounce/50-gram ramekins

Vegetable Confit

Restaurant menus often use the word *confit* to describe vegetables. This takes some liberty with the term but it seems to have become part of our culinary vernacular, and, as it suggests a product that is very tender and very rich, it's not entirely inappropriate to call eggplant diced and cooked to a virtual puree eggplant confit.

Here are two vegetable "confits" that have myriad uses. They are distinguished mainly by their seasonings, since the basic method for preparing any confit is to heat low and slow till the meat is all but melting. Vegetable confits can be served warm or cold, by themselves as you would a chutney or added to sauces or other dishes to enhance them. They're like vegetable concentrates and could even be added to pâtés as a seasoning.

ONION CONFIT

This onion confit is cooked-down onions with a little seasoning and, to balance the intense sweetness of the onions, a little acid. But it demonstrates exactly how rich and complex a plain old onion really is. The confit can be used as an accompaniment for a terrine or it can be added to other sauces, such as the Caraway-Beer Mustard (page 286), for additional complex sweetness. It goes well with smoked items and is perfect with sausages. You could even spread it on a roast beef sandwich and it would be delicious.

> 3 ounces/80 grams unsalted butter
> 2 pounds/1 kilogram Spanish or other sweet onions, halved
> lengthwise and thinly sliced
> ½ cup/125 milliliters dry white wine
> ¼ cup/60 milliliters honey
> ¼ cup/60 milliliters white wine vinegar
> Kosher salt and freshly ground black pepper

1. Melt the butter in a large heavy-bottomed sauté pan over medium heat. Add the onions and cover. Cook until soft, approximately 30 to 45 minutes, stirring occasionally.

2. Remove the lid and season the onions with salt and pepper. Add the wine, honey, and vinegar, turn the heat to medium-high, and cook until most of the liquid has evaporated. Remove from the heat and let cool.

3. Chill before serving. The onion confit will keep for 3 weeks, tightly covered, in the refrigerator.

Yield: 2 cups/500 milliliters

TOMATO CONFIT

Like onion confit, this can be used to accompany pâtés and terrines, and it also works well with smoked meats and sausages.

> 2 tablespoons/20 grams minced or grated peeled fresh ginger
> 2 tablespoons/36 grams minced shallots
> 1 ½ tablespoons/28 grams minced garlic
> ¼ cup/60 milliliters extra virgin olive oil
> 1 ½ tablespoons/25 milliliters red wine vinegar
> 6 ripe Roma (plum) tomatoes, peeled, seeded,
>   and diced
> ¼ cup/25 grams minced fresh herbs, such as flat-leaf basil,
>   parsley, and chives, or a combination
> Kosher salt and freshly ground black pepper

1. Combine the ginger, shallots, garlic, oil, and vinegar in a deep skillet large enough to contain the tomatoes and bring to a boil. Reduce the heat to medium, add the tomatoes, and cook for 12 to 15 minutes until soft and paste-like.

2. Transfer to a bowl to cool, then fold in the herbs and season with salt and pepper. The confit keeps for a week in the refrigerator. Serve warm or at room temperature.

Yield: 2 cups/500 milliliters

8.

SAUCES AND CONDIMENTS:

NOT OPTIONAL

. . . .

Most chefs worth their salt would as willingly enter the dining room naked as send a dish to a customer without a sauce. Sauces and condiments should not be thought of, in the professional kitchen or at home, as extras or add-ons, but rather as fundamental parts of any given dish, just as seasoning is. They're also fun and easy to make. Brian and I like to put out several sauces, chutneys, and relishes with a variety of sausages or terrines, because they're so visually compelling. The following recipes use basic techniques–preparing a vinaigrette, a mayonnaise, a chutney–to build great sauces and condiments that work especially well in the charcutier's kitchen.

Condiments and sauces are essential for most charcuterie items—they finish a plate, complete it, adding the final seasoning, acidity or sweetness, moisture, richness, and visual and textural elements, no matter whether a chunky salsa, a refined reduction, or a flavored mayonnaise. Among the first information culinary students learn is the unspoken rule that no dish, from beginning of the meal to the end, is served without an accompanying sauce of some sort—one reason why the saucier is a position of high esteem in a classic French kitchen. Sauce is what makes a dish excellent rather than good. A sausage may be tasty, but it's better with some mustard. A plain slice of veal terrine is fine, but spoon a pool of orange-ginger sauce beside it, and it's complete. When you've taken the time to craft a terrine, the specialness of that preparation all but demands an equally cared-for sauce.

Brian and I have tried to give a variety of all-purpose sauces and condiments—from chutneys to mustards to mayonnaises and vinaigrettes—as well as sides such as potato salad and pickle chips, that happen to go especially well with the food in this book, from a fancy pâté en croûte to a rustic smoked pork shoulder.

When deciding on a type of sauce, ask yourself how you want to elevate the dish. Should the sauce be sharp like a mustard or a vinaigrette, or should it be rich and creamy like an aïoli? Should it be salty or sweet, thick or thin, with fresh herbs or without? We've made certain recommendations for which foods and sauces to pair, but most sauces are infinitely adaptable to variations, depending on the cook's mood and the particular nuances of any given dish. And though we've made this the last chapter in the book, sauces should not be an afterthought, or put together only if there's extra time when the real cooking's done. Sauces are fundamental to excellence in the kitchen and at the table.

BASIC MAYONNAISE

Mayonnaise is one of the great all-purpose sauces. Americans tend to think of it in terms of a sandwich spread or something to turn cold starch items like boiled potatoes or macaroni into salads. But at heart it's a sauce that can be varied for just about anything, with a few additions: for pork, for example, try adding cumin, cayenne, and lime juice; for fish, saffron and garlic; for chicken, lemon juice and tarragon; and for beef, just horseradish. If you need to make an elegant sauce in thirty seconds, mayonnaise is the way to go.

Anyone who likes to cook should be able to make a good mayonnaise. Hellmann's is a decent everyday mayonnaise and fine when making a mayo-based sauce on the fly or for slathering on a BLT or fried egg sandwich. In Michigan, the jarred sauce of choice seems to be Miracle Whip, a sweeter-style mayo that we find harder to recommend. Making your own allows you to tailor it to your own tastes, and when you use a fresh neutral oil, such as a good vegetable oil or canola oil or grapeseed oil, and fresh seasoning, the result is far better both in flavor and in texture than grocery-store mayonnaise.

We recommend using yolks only from organic eggs, or of eggs from known local farmers, when making mayonnaise. We've never had a problem with bacterial contamination. It is still a possibility, however, so it's prudent to avoid feeding infants or the infirm raw egg products.

Mayonnaise can be made using a whisk or with a mortar and pestle, if it's the right size and shape–the traditional method for making aïoli. Mayo made in a mortar and pestle is as smooth and rich as face cream. Make your mayonnaise this way, and you too can develop forearms like Popeye. For quantities larger than a cup, a food processor or a standing mixer will work. Any way you choose to achieve a stable emulsion is fine.

The basic ratio for mayonnaise is one yolk per cup (250 milliliters) of oil. The key is to begin very slowly. This kind of emulsion is created by breaking up the oil into minuscule droplets that remain separated by water via an emulsifier called lecithin (found in egg yolks). Some believe that a stable emulsion is easier to achieve when both eggs and oil are at room temperature. If the droplets touch and recombine, the emulsion falls apart, or breaks, and you've got soup. If you sense this is about to happen, add another teaspoon of water. If your rich stiff emulsion suddenly liquefies before your eyes, you'll need to begin with a fresh yolk and fresh oil. Once you've got a stable emulsion going, you can add your broken one as if it were oil, in a thin, slow stream.

The critical time is the first addition of oil, when the oil should be added a few drops at a time. Once the emulsion has been established, the oil can be added in a thin stream, still whisking or processing continuously, till all the oil is incorporated and emulsified.

Properly seasoned mayonnaise is one of the great all-purpose sauces and one of the most versatile as well.

1 large egg yolk, preferably organic
½ teaspoon/4 grams kosher salt
Pinch of white pepper
1 teaspoon/5 milliliters water
1 to 2 teaspoons/5 to 10 milliliters fresh lemon juice, to taste
1 cup/250 milliliters vegetable oil

1. Combine the yolk, salt, pepper, water, and a few drops of the lemon juice in a medium bowl. Fold a dishtowel into a ring on the counter and set the bowl in this ring to keep it steady, and whisk the ingredients together.

2. It helps to measure out your oil into a cup that pours well in a wire-thin stream; alternatively, you can start your emulsion by drizzling the oil off a spoon, then slowly pouring in the rest of the oil after the emulsion has begun. Add the oil slowly while whisking vigorously: start by adding just a few drops of oil as you whisk; when the emulsion becomes creamy, increase the speed with which you add the oil to a thin stream. Once the emulsion was started, the mixture should be thick enough to hold its shape and look luxuriously creamy. When all the oil is incorporated, add additional lemon juice to taste. If the mayonnaise is too thick, thin it by whisking in a little water.

Yield: About 1 cup/250 milliliters

AÏOLI

This mayo variant is a staple in Provence and other southern European regions—flavorful olive oil replaces the neutral oil of the mayonnaise and garlic is added. Aïoli goes with practically anything, and variations are, as with mayonnaise, limited only by your imagination. Minced or pureed herbs are an easy and colorful addition. Add pureed roasted red pepper to make rouille, a traditional Mediterranean accompaniment to fish stews. Whisking in a half-teaspoon of hot water infused with saffron turns this into an extraordinary sauce for fish (whether whole or in a terrine or sausage). Aïoli goes beautifully with potatoes, crudités, and toasted baguette slices, and it can be used, as rouille is, to enrich stews and soups.

> 1 large egg yolk, preferably organic
> 1½ teaspoons/8 milliliters water
> ½ teaspoon/2.5 milliliters white wine vinegar
> 1 teaspoon/7 grams kosher salt, or to taste
> 1½ tablespoons/22.5 milliliters fresh lemon juice, or to taste
> 1 teaspoon/6 grams garlic minced and smashed to a paste
> 1 cup/250 milliliters extra virgin olive oil

1. Combine the egg yolk, water, vinegar, salt, a few drops of lemon juice, and garlic in a medium bowl and whisk to combine.

2. Whisking continuously, drizzle in the oil, drop by drop at first, then in a very thin stream, until oil is incorporated. Taste and adjust the seasoning with lemon juice and/or salt as necessary. Thin the aïoli with a few drops of water if it's too thick.

Yield: About 1 cup/250 milliliters

RÉMOULADE

A classic French sauce all too rarely used other than with grated celery root in the traditional *céleri rémoulade*, perhaps because of its bastardization in the form of tartar sauce, as in fish sticks with tartar sauce. But it remains an excellent sauce, similar to Sauce Gribiche (page 282), mayo-based with lots of chopped tart or salty garnish, for any cold or warm fish preparation and, with its high-fat, high-acid content, is especially good with seafood terrines.

2 cups/500 milliliters Basic Mayonnaise (see page 279)

1 tablespoon/15 milliliters Dijon mustard

¼ cup/60 grams finely chopped cornichons

1 tablespoon/15 grams chopped capers

1 anchovy fillet, finely chopped

3 tablespoons/24 grams chopped fresh flat-leaf parsley

1 teaspoon/3 grams chopped fresh chervil (optional)

1 teaspoon/3 grams chopped fresh tarragon

1. Put the mayonnaise and mustard in a bowl, and fold all the remaining ingredients.
2. Store in the refrigerator in an airtight container for up to a week.

Yield: 2 ½ cups/625 milliliters

SAUCE GRIBICHE

Sauce gribiche is a classic French sauce using cooked egg yolks, and it typically includes small tart pickles (gherkins or cornichons) and lots of herbs, notably tarragon. It's a delicious tangy vinaigrette-mayo hybrid. There are all kinds of variations, some loose, others emulsified. This version is straight out of *Escoffier: The Complete Guide to the Art of Modern Cookery*, published in 1907. It's fully emulsified, the consistency of mayonnaise. Escoffier recommended serving it with fish terrines. In fact, it goes equally well with pork dishes and may be the perfect all-purpose charcuterie sauce.

3 large hard-cooked eggs

1 ½ teaspoons/7.5 milliliters Dijon mustard

1 tablespoon/15 milliliters fresh lemon juice, or to taste

Kosher salt

2 teaspoons/10 milliliters white wine vinegar

¾ cup/185 milliliters vegetable oil or canola oil

1 ½ tablespoons/35 grams chopped capers

3 tablespoons/45 grams chopped cornichons

1 teaspoon/3 grams minced fresh flat-leaf parsley

1 teaspoon/3 grams minced fresh tarragon

1 teaspoon/3 grams minced fresh chervil

Freshly ground black pepper

1. Separate the egg yolks from the whites. Press the yolks through a fine sieve into a medium bowl. Add the mustard, lemon juice, and vinegar and season with a healthy pinch of salt. Whisking continuously, drizzle in the oil, drop by drop at first, then in a very thin stream, until all the oil is incorporated and you have a smooth, thick emulsion.

2. Finely dice the egg whites and fold them into the sauce, along with the remaining ingredients. Taste and adjust the seasoning, adding more lemon juice and/or salt and pepper as necessary.

Yield: 1 cup/250 milliliters

CUCUMBER DILL RELISH

This sour cream–based sauce is perfect for seafood terrines and seafood sausages. It's also excellent for any salmon, whether cured, smoked, or poached.

1 English cucumber, peeled and sliced into $1/4$-inch/
0.5-centimeter disks
1 tablespoon/15 grams kosher salt
3 tablespoons/45 milliliters red wine vinegar
$1/2$ cup/70 grams minced Spanish onion
3 tablespoons/24 grams chopped fresh dill
1 teaspoon/5 grams sugar
$3/4$ cup/185 milliliters sour cream
$1/4$ cup/60 milliliters extra virgin olive oil
Freshly ground black pepper to taste

1. Toss the cucumbers with the salt and a splash of the vinegar in a bowl, and set aside to marinate for 1 hour.

2. Drain the cucumber and dice. Toss gently with the remaining vinegar and other ingredients except the pepper. Taste for seasoning, then add the pepper.

3. Cover and allow to stand in the refrigerator for 2 to 3 hours, and up to a day, before serving.

Yield: 2 cups/500 milliliters

SMOKED TOMATO AND CORN SALSA

This is a colorful mix of chopped vegetables and aromatics, a straightforward pico de gallo–style sauce. The smokiness of the tomatoes can go with smoked or nonsmoked items. The lime, chiles, and cilantro give the salsa a sharp Southwestern flavor that pairs well with fish, poultry, and pork. This recipe makes a lot, and it can easily be halved–but a bowl of it next to some corn chips disappears pretty quickly.

6 Roma (plum) tomatoes
2 ears fresh corn on the cob, shucked
Kosher salt and pepper to taste
1 tablespoon/18 grams minced garlic
½ teaspoon/1 gram minced lemon zest
Juice of 1 lime
¼ cup/60 milliliters olive oil
2 serrano chiles, seeded and minced (substitute jalapeños
if necessary)
2 tablespoons/16 grams chopped fresh cilantro
3 tablespoons/24 grams chopped fresh mint

1. Cut the tomatoes in half and cold-smoke (see page 74) for 2 to 3 hours.

2. Season corn with salt and pepper, rub with a little oil, and grill over an open flame until golden brown. Cook. Cut the kernels off the cob using a sharp knife.

3. Place the tomatoes in a food processor with the garlic and pulse until coarsely chopped. Transfer to a bowl, add the corn and the remaining ingredients, and stir gently to mix. Cover and let stand for at least 30 minutes before serving.

Yield: 3 cups/750 milliliters

Three Mustards

These three mustards, each very different in flavor, show off mustard's versatility. Paired with a tart fruit, a spicy vegetable, or an unusual aromatic, it can be used variously as a sauce for a single sausage or a plate of sliced sausages, as well as for pâtés. Each uses the same technique: egg yolks, liquid, dry mustard, and the main flavoring ingredients are cooked over simmering water till thickened, then cooled before serving. These will keep for a week to ten days refrigerated in an airtight container.

TART CHERRY MUSTARD

Brian created this tart fruit mustard to make use of one of Michigan's outstanding natural products. It goes well with pork and sausages, especially Teutonic ones such as Thuringer, hunter, and summer sausages.

2 tablespoons/18 grams Colman's dry mustard

3 large egg yolks

½ cup/125 milliliters cranberry juice

3 tablespoons/45 milliliters white wine vinegar

¼ teaspoon/2 grams salt, or more to taste

½ tablespoon/10 milliliters Worcestershire sauce

2 tablespoons/30 grams light brown sugar

Pinch of cayenne pepper

3 tablespoons/50 grams finely chopped dried tart cherries

1. Combine all the ingredients except the cherries in a metal bowl or the top of a double boiler and cook over simmering water, whisking continuously but gently until thickened and smooth, about 15 minutes; a whisk drawn through the mustard should leave a line. Do not whisk it too vigorously, or it will become frothy. Remove from the heat and add the cherries. Taste and adjust the seasoning.

2. Refrigerate, covered, until chilled.

Yield: 1 cup/250 milliliters

GREEN CHILE MUSTARD

Flavored with roasted green chile peppers and tequila, this mustard goes particularly well with smoked chicken sausages, duck sausage, and pork terrines.

1 poblano chile

1 jalapeño chile

1 tablespoon/8 grams cumin seeds

1 tablespoon/8 grams pure chile powder

1 tablespoon/9 grams Colman's dry mustard

¼ cup/60 milliliters malt vinegar

¾ cup/185 milliliters lager beer
¼ cup/185 milliliters honey
3 tablespoons/45 milliliters tequila
4 large egg yolks

1. Roast the chiles over an open flame or under a broiler, turning frequently, until the skin turns black and blisters. Transfer to a bowl, cover with plastic wrap, and let cool.

2. Peel the chile peppers, cut them in half, remove the seeds, and mince.

3. In a small dry sauté pan, toast the cumin seeds and chili powder briefly over medium-high heat, just until they release their fragrance, about 2 minutes (be careful not to burn them).

4. Transfer the toasted spices to a metal bowl or the top of a double boiler and add the chiles and the remaining ingredients. Cook over simmering water, whisking continuously but gently, approximately 15 minutes, until smooth and thickened, just slightly looser than a mayonnaise. Do not whisk too vigorously or it will become frothy. Remove from the heat.

5. Refrigerate, covered, until chilled.

Yield: About 2 cups/500 milliliters

CARAWAY-BEER MUSTARD

One of the oldest cultivated spice plants in the West, caraway is now used mainly in German cooking, notably in breads and pastries, as well as in cheeses and pickles and braised dishes. Combining as it does the flavors of beer, caraway, and mustard, this is, not surprisingly, a perfect sauce to serve with German sausages such as summer sausage or Thuringer.

2 tablespoons/18 grams Colman's dry mustard
6 tablespoons/85 milliliters beer
1 ½ teaspoons/10 milliliters Worcestershire sauce
1 tablespoon/8 grams caraway seeds, toasted and crushed
  (see Note page 51)
3 ounces/85 milliliters malt vinegar
2 tablespoons/30 milliliters cup honey

½ teaspoon/5 grams kosher salt

3 large egg yolks

½ teaspoon/5 grams sugar

1. Combine all the ingredients in a metal bowl or the top of a double boiler. Cook over simmering water, whipping continuously but gently, until thickened. Do not whisk too vigorously, or it will become frothy. Remove from the heat.

2. Refrigerate, covered, until chilled.

Yield: About 1 cup/250 milliliters

BASIC VINAIGRETTE

A straightforward vinaigrette is an essential element of any cook's repertoire. The variations are endless, but the ratio is the same: generally, three parts oil to one part vinegar (lemon and lime juices are more acidic and require four parts oil). The rest is seasoning. This recipe stays true to the classical tradition with shallot and Dijon mustard for the seasonings (the mustard also helps to maintain the emulsification), but this is a matter of taste. Minced garlic is excellent in a vinaigrette—a simple lemon juice, garlic, and black pepper vinaigrette, for example. Vary the type of the oil, if you wish. The vinegar can be red wine, white wine, champagne, or sherry. Balsamic vinegar can also be a great component of a vinaigrette, but because of its intense sweetness, use it as a seasoning element rather than a main ingredient. Or use another form of acid: strained fresh citrus juices or verjus (the juice of unripened wine grapes), perhaps. You can also vary your vinaigrette according to what it's dressing. If you're serving a beet and walnut salad, for instance, you might use one part walnut oil and two parts canola oil. For a duck confit salad, some of the confit jelly (see page 255) can be substituted for some of the salt and oil. For a black bean salad, try lime juice and cumin with a neutral oil.

Fresh herbs are excellent in almost any vinaigrette, but they should be added at the last minute or the acid will discolor them.

Vinaigrette technique is simple and again, in part a matter of preference. The vinegar, salt, and shallots should be combined first so that the salt dissolves completely (it's less evenly distributed when added to fat or oil), and so that the flavor of the shallots softens and is evenly distributed. But how you mix the vinaigrette is up to you. It can be done in a blender (though in that case it can become so thick that you may have to transfer the vinaigrette to a bowl and finish whisking by hand), a food processor, or in a bowl with a whisk; you can even throw it all into a jar and shake it. The result, of course, differs—simply shaken, it will quickly separate. No matter which method you use, the vinaigrette should be emulsified to the consistency of a pourable mayonnaise. If it's soupy, it will eventually separate. (This alters the way it coats the greens but not the way it tastes.)

The uses of a vinaigrette are countless. Some chefs consider it to be a "mother sauce" because of the range of variations possible with the basic formula. With only small variations in seasonings, oils, and acidity levels, a vinaigrette can be excellent with almost any food—meat, fish, chicken, vegetables, legumes, potatoes, what you will.

In fact, a vinaigrette is probably the most important and powerful sauce available to any cook, and also the easiest to make. But use good-quality vinegars and oils. That will make all the difference.

½ cup/125 milliliters vinegar (red wine, white wine, sherry
     or champagne)
1 tablespoon/18 grams minced shallots washed under cold
     running water
¼ teaspoon/2 grams kosher salt, or to taste
Freshly ground black pepper to taste
2 tablespoons/30 milliliters Dijon mustard
1 ½ cups/375 milliliters oil (vegetable, canola, grapeseed,
     corn, olive)

1. Combine the vinegar, shallots, salt, then the pepper and mustard, in a medium bowl. Whisk in the oil, drop by drop at first, then in a thin, steady stream.

2. Refrigerate until ready to use. (A vinaigrette can be refrigerated for 1 to 2 weeks.)

Yield: 2 cups/500 milliliters

RUSSIAN DRESSING

This is an all-but-forgotten tomato-mayo–based sauce that's ripe for a come-back—it's a real treasure. Of course, it's what makes a Reuben what it is, but it's excellent with so many items beyond cured beef and chef's salad—elevating vegetables, eggs, and smoked white meats.

> 1 cup/250 milliliters Basic Mayonnaise (page 279)
> 1 tablespoon/10 grams prepared horseradish, squeezed dry
> ¼ cup/60 milliliters chili sauce
> 1 tablespoon/18 grams minced onion
> 1 teaspoon/5 milliliters Worcestershire sauce
> Kosher salt and freshly ground black pepper to taste

Put the mayonnaise in a bowl and fold in all the remaining ingredients. The dressing will keep for 1 week in the refrigerator in a container with tight-fitting lid.

Yield: 1¼ cups/300 milliliters

CHIPOTLE BARBECUE SAUCE

This is an intense, complex sauce, very sweet and sour, heavily spiced, with the distinctive flavor and heat of the smoked jalapeño pepper, known as the chipotle. This is a delicious all-purpose barbecue sauce, especially good with slow-roasted pork shoulder.

> ½ cup/70 grams chopped onion
> 1 tablespoon/18 grams chopped garlic
> 1 tablespoon/15 milliliters olive oil
> ½ cup/125 milliliters cider vinegar
> ½ teaspoon/2 grams ground cloves
> ½ teaspoon/2 grams ground coriander
> ¼ teaspoon/1 gram ground allspice
> ¼ packed cup/50 grams dark brown sugar
> 1 tablespoon/15 milliliters molasses

1 tablespoon/16 grams seeded and minced chipotle peppers
in adobo sauce
¾ cup/185 milliliters catsup
1 tablespoon/15 milliliters Worcestershire sauce
½ cup/125 milliliters veal stock, chicken stock, or water

1. Sauté the onion and garlic in the oil in a medium heavy-bottomed saucepan over medium heat until well browned, about 15 minutes.

2. Add the vinegar and simmer to reduce by half. Add the remaining ingredients, bring to a gentle simmer, and cook for 15 minutes.

3. Puree with an immersion blender, or transfer to a regular blender and puree. The sauce keeps for a week refrigerated.

Yield: 1½ cups/375 milliliters

CAROLINA-STYLE BARBECUE SAUCE

The term *barbecue* refers to a specific preparation in the Carolinas, often called pig pick, pulled pork, a slow-cooked pork shoulder tossed with a vinegar-based sauce, sometimes with tomato, served on a bun or beside a pile of hushpuppies. In addition to being a component of Carolina barbecue, this sauce would work well with braised chicken, duck, or beef. But it's best with North Carolina hog.

½ cup/70 grams minced onion
1 tablespoon/18 grams minced garlic
1 tablespoon/15 milliliters vegetable oil
½ cup/125 milliliters cider vinegar
½ cup/125 milliliters Worcestershire sauce
1 tablespoon/4 grams Colman's dry mustard
2 tablespoons/26 grams dark brown sugar
2 tablespoons/16 grams paprika
1 tablespoon/15 grams kosher salt
1 teaspoon/3 grams cayenne pepper
1 cup/250 milliliters catsup

1. In a medium heavy-bottomed saucepan, gently sauté the onion and garlic in the oil until softened but not colored.

2. Add all the remaining ingredients except the catsup and bring to a simmer. Stir in the catsup and cook gently for 15 minutes. Remove from the heat. The sauce keeps for up to 1 week refrigerated.

Yield: 2 cups/500 milliliters

CUMBERLAND SAUCE

This classic sauce, with a dominant profile of red currant and mustard, is based on Escoffier's recipe and is ideal for rich venison terrines, or any rich duck or goose or pork dish. It's often served cold, so the fruity, spicy, acidy components are used aggressively.

Grated zest and juice of 1 orange
Grated zest and juice of 1 lemon
2 tablespoons/36 grams minced shallots
1 ¼ cups/300 milliliters currant jelly
1 ½ teaspoons/4 grams Colman's dry mustard
¾ cup/185 milliliters ruby port
½ teaspoon/3 grams kosher salt
Pinch of cayenne pepper
Pinch of ground ginger

1. Blanch the zest and shallots in boiling water for 45 seconds. Drain in a strainer and cool under cold running water; drain again.

2. Transfer to a large heavy-bottomed saucepan, add all the remaining ingredients, and bring to a simmer. Simmer until reduced to the consistency of a light syrup, about 15 minutes. Strain the sauce through a fine-mesh strainer, and refrigerate until chilled.

Yield: About 1 cup/250 milliliters

ORANGE-GINGER SAUCE

This sauce uses a similar flavor principle as the Cumberland sauce, combining fruit and pepper for an aggressively flavored sweet-sour-spicy sauce. The uncooked alcohol, sherry, gives it a strong aromatic presence. Orange and ginger, always an excellent pair, go well with pork, duck, goose, and venison dishes.

> 1 tablespoon/4 grams grated orange zest
> 1 tablespoon/4 grams grated lemon zest
> 2 teaspoons/8 grams grated fresh ginger
> 1 cup/250 milliliters orange marmalade
> ¼ cup/50 grams dark raisins
> 2 tablespoons/30 milliliters fresh lemon juice
> 2 tablespoons/30 milliliters dry sherry
> 1 teaspoon/3 grams ground ginger
> 1 tablespoon/15 milliliters honey
> ½ teaspoon/1 gram Colman's dry mustard
> ⅛ teaspoon/1 gram cayenne pepper

Combine all the ingredients in a food processor and process until smooth. Store, covered, in the refrigerator.

Yield: 1½ cups/375 milliliters

HORSERADISH CREAM SAUCE

This is a straightforward sour cream and mayonnaise sauce featuring horseradish, a root that ought to be used more in the home kitchen. It's best to use a Microplane to grate it, if you have one. While prepared horseradish can be substituted, freshly grated horseradish root is superior because it's not obscured by the vinegar and salt in the bottled version. This sauce is excellent with smoked pork, chicken, and turkey, not to mention grilled beef and beef sandwiches. It also happens to work extremely well with Brian's Holiday Kielbasa (page 117).

2 ounces/60 grams fresh horseradish, peeled
and finely grated
½ cup/125 milliliters Basic Mayonnaise (page 279)
½ cup/125 milliliters sour cream
2 teaspoons/10 milliliters fresh lemon juice
Kosher salt and ground white pepper to taste

Combine all the ingredients in a bowl or other container. Store, covered, in the refrigerator.

Yield: About 1 cup/250 milliliters

BASIL CREAM SAUCE

Here are two versions of an elegant and easy sauce that can accompany many seafood, vegetable, and veal terrines and sausages. Serving a sliced seafood sausage in a circle of basil cream sauce elevates it to another level. The first of these (mayonnaise-based) is incredibly easy, the second (cream-based) is simply easy.

Basil Cream Sauce I

1 cup/250 milliliters Basic Mayonnaise (page 279)
½ cup/50 grams fresh basil leaves
1 tablespoon/15 milliliters white wine vinegar
Kosher salt and ground white pepper to taste

Combine all the ingredients in a blender and blend until the basil is finely pureed and evenly mixed throughout. If the mayonnaise is too thick to blend well, add a tablespoon of water.

Yield: 1 cup/250 milliliters

Basil Cream Sauce II

> 1 cup/100 grams fresh basil leaves
> 1 cup/250 milliliters heavy cream
> Kosher salt and ground white pepper to taste

1. Blanch the basil in a saucepan of heavily salted water for 10 seconds, then transfer it to an ice bath to cool. Drain, and squeeze out the excess water.

2. Bring the cream to a boil. Reduce over medium-high heat until thick. Remove from the heat, let cool, and transfer to a blender. Add the basil and salt and pepper and blend until the basil is completely pureed, about 2 minutes. Taste for seasoning, and adjust if necessary.

3. Strain the sauce through a fine-mesh strainer and refrigerate until chilled.

Yield: About 1 cup/250 milliliters

SPICY TOMATO CHUTNEY

This is a highly spiced sauce of fresh tomatoes cooked down to a thick consistency. It can be served hot, warm, or cold with smoked, grilled, or roasted meats.

> 4 large Roma (plum) tomatoes (about 1 pound/450 grams),
>   peeled, seeded, and roughly chopped
> 1 tablespoon/10 grams grated fresh ginger
> ½ teaspoon/4 grams coriander seeds, toasted and ground
>   (see Note page 51)
> ⅛ teaspoon/1 gram cayenne pepper
> ½ teaspoon/2 grams ground turmeric
> ⅛ teaspoon/1 gram ground cardamom
> ½ teaspoon/2 grams Colman's dry mustard
> ½ tablespoon/9 grams minced garlic
> ¼ cup/60 milliliters cider vinegar
> 2 tablespoons/30 milliliters honey

1. Combine all the ingredients in a heavy-bottomed saucepan and gently cook over medium heat until reduced and thickened to a chutney-like consistency, about 20 minutes.

2. Cool, then cover and refrigerate.

Yield: 1 cup/250 milliliters

CORN RELISH

Great color, vibrant on the plate, this relish works with virtually any sausage and meat terrine. Brian and I tend to make it September through October, when the corn has become a little more sweet.

6 cups/600 grams fresh corn kernels

1 red bell pepper, seeded and diced

1 green bell pepper, cured, seeded, and diced

½ cup/70 grams onion, diced small

4 small Roma (plum) tomatos (8 ounces/225 grams) peeled,
    seeded, and diced

1 cup/225 grams sugar

1 ½ cups/375 milliliters cider vinegar

1 ½ teaspoons/8 grams kosher salt

¾ teaspoon/3 grams ground white pepper

1 teaspoon/3 grams ground turmeric

1 teaspoon/4 grams mustard seeds

2 tablespoons/20 grams cornstarch

2 tablespoons/15 milliliters water

1. Combine all the ingredients except the cornstarch and water in a nonreactive saucepan and bring to a boil. Reduce the heat to low and simmer gently for 30 minutes, stirring frequently.

2. Dissolve the cornstarch in the water, and stir into the relish to thicken it. Remove from the heat.

3. Cool, then cover and refrigerate.

Yield: 1 quart/1 liter

GREEN TOMATO RELISH

This is an early season relish, when you can't wait for the tomatoes to ripen, or a late-summer sauce, when you have a bounty of unripe tomatoes. Its sweet-and-sour flavor goes best with pork dishes–terrines, confit, smoked pork.

3 juniper berries, smashed with the side of a knife

6 black peppercorns, cracked beneath a small heavy pan or
    with the side of a knife

2 whole cloves

2 large green tomatoes, cut into large dice (about 2 cups/
    400 grams)

1 cup/140 grams onion, sliced thin

1 cup/100 grams diced (about 1 inch/2.5 centimeters)
    unpeeled Granny Smith apple

¼ cup/50 grams golden raisins

2 tablespoons/30 milliliters cider vinegar

1 tablespoon/10 grams grated fresh ginger

⅓ packed cup/75 grams light brown sugar

2 teaspoons/10 grams minced garlic

¼ cup/60 milliliters chicken stock

1. Tie the juniper, peppercorns, and cloves in a square of cheesecloth to make a sachet (a coffee filter will also work).

2. Combine the sachet with the remaining ingredients in a large heavy-bottomed saucepan and bring to a boil over high heat. Reduce the heat to low and simmer until reduced and thickened, about 45 minutes.

3. Taste and adjust the seasonings if necessary. Discard the sachet, and store the relish covered in the refrigerator.

Yield: 2 cups/500 milliliters

ONION-RAISIN CHUTNEY

This is a traditional chutney and is particularly good with the more refined pork, veal, or chicken terrines, those with fine interior garnish or inlays.

2 cups/200 grams diced onions
1 cup/200 grams dark raisins
½ cup/125 milliliters cider vinegar
¼ packed cup/75 grams light brown sugar
½ teaspoon/2 grams ground turmeric
1 cinnamon stick
1 bay leaf
Pinch of ground allspice
Kosher salt and freshly ground black pepper to taste

1. Combine all the ingredients in a heavy-bottomed saucepan, bring to a simmer, and simmer for 20 to 30 minutes, stirring often, until the juices are thick and syrupy.
2. Remove the cinnamon and bay leaf. Cool, then store, covered, in the refrigerator.

Yield: 2 cups/500 milliliters

BOURBON GLAZE

This glaze can be used on any smoked poultry or ham, or even on roasted meats, for an excellent color and spicy-sweet coating. Smoke or roast the meat until it begins to get color on its own, then brush the meat with this glaze once or twice as it finishes cooking and once again when it's off (or out of) the heat.

1 cup/250 milliliters whiskey, bourbon, or Wild Turkey
½ cup/125 grams maple sugar or ½ cup/125 milliliters maple syrup
¼ packed cup/50 grams dark brown sugar
Pinch of cayenne pepper

1. Combine all the ingredients in a heavy-bottomed pot and bring to a simmer, then

lower the heat and simmer gently until the glaze has reduced to 1 cup/250 milliliters and has a syrupy consistency, about 15 minutes.

2. Remove from the heat and let cool, then refrigerate, covered, until ready to use.

Yield: 1 cup/250 milliliters

MARINATED OLIVES

Marinated olives are an excellent accompaniment to a plate of salami and good bread. Store-bought olives come in an oil or brine that almost always can be greatly improved upon. This marinade includes herbs, garlic, and citrus peel. Use a variety of brined and oil-cured olives–Kalamata, Niçoise, Picholine, Moroccan, Sicilian (today even grocery stores often have a good selection)–for diverse flavors and a dynamic visual appeal.

1 pound/450 grams assorted brined and oil-cured olives
3 garlic cloves, sliced paper-thin
¼ cup/24 grams chopped fresh oregano
¼ cup/24 grams chopped fresh flat-leaf parsley
1 tablespoon/8 grams fresh thyme leaves
2 strips orange peel, about 2 inches/5 centimeters long and
        ¼ inch/0.5 centimeter wide, white pith removed
2 strips lemon peel, about 2 inches/5 centimeters long and
        ¼ inch/0.5 centimeter wide, pith removed
1 tablespoon/15 milliliters fresh lemon juice
Extra virgin olive oil to cover

1. Drain the olives and rinse well under warm running water. Dry the olives on a towel.

2. Crack about one-third of the olives, under a small heavy skillet or the side of a knife. Remove the pits from these, or all the olives, as desired.

3. Place the olives in a bowl, cover with olive oil, add the remaining ingredients, and toss. Transfer to a container with a tight-fitting lid and allow to marinate for at least 2 days in the refrigerator before serving.

Yield: About 1 pound/450 grams

GERMAN POTATO SALAD

With its richness and acidity, potato salad is a perfect, and traditional, accompaniment to sausages and terrines. It tastes great cold or warm, and is extremely versatile. A small amount of good potato salad will turn a slice of terrine into a first course. The key is to have the right acidity and salt levels, and to cook the potatoes properly: whole, with the skin on, in heavily salted water barely at a boil (you don't want the skin to split as that would waterlog the potatoes). After that, variations are unlimited. The following recipe is classical with its flavoring of bacon, mustard, vinegar, and chives.

2 ¼ pounds/1 kilogram Yukon Gold potatoes
⅓ cup/100 grams lardons (slab bacon cut into batons;
      see page 40)
½ cup/70 grams diced onions
¼ cup/60 milliliters vegetable oil
1 cup/250 milliliters Chicken Stock (see page 226)
¼ cup/60 milliliters white wine vinegar
1 teaspoon/5 grams sugar
2 tablespoons/30 milliliters whole-grain mustard
¼ cup/32 grams chopped fresh chives
Kosher salt and ground white pepper
Pinch of cayenne pepper

1. Gently simmer the potatoes in heavily salted water just until tender (a skewer or paring knife should pass through a potato without resistance, but they should not be falling apart). Drain the potatoes in a colander and let excess water steam off.

2. Meanwhile, prepare the dressing: Sauté the lardons in a small sauté pan over medium-low heat until crisp on the outside but still soft inside. Transfer the lardons to a plate to cool, and reserve the rendered fat.

3. In a medium saucepan, cook the onions in the vegetable oil until soft but not colored. Add the chicken stock, vinegar, and sugar and bring to a boil, then remove from the heat.

4. As soon as the potatoes are cool enough to handle—they should still be warm—peel the potatoes and slice into ¼-inch/0.5-centimeter slices; try to keep them uniform. Transfer to a bowl.

5. Add the reserved bacon fat and the mustard to the warm potatoes. Slowly stir in the onion mixture as needed (very moist but not soupy). Gently fold in the lardons and chives, keeping the potatoes intact. Season with salt, black pepper, and cayenne and gently toss.

6. Serve warm, at room temperature, or chilled.

Yield: 10 servings

SWEET PICKLE CHIPS

These are a simple and excellent accompaniment to any meat sausage or rustic terrine—thickly sliced cucumber cooked in acid and spices to make sweet, sour, crunchy pickle chips.

1 pound/450 grams cucumbers, sliced into ¼-inch/
0.5-centimeter disks
¼ pound/115 grams sliced white onions
¾ cup/185 milliliters cider vinegar
1½ teaspoons/8 grams kosher salt
¼ teaspoon/2 grams mustard seeds
1 cup/225 grams sugar
1 teaspoon/4 grams celery seeds
½ cup plus 2 tablespoons/125 milliliters plus 30 milliliters
white wine vinegar
1 teaspoon/4 grams allspice berries
½ teaspoon/2 grams ground turmeric

1. Combine the cucumbers, onions, cider vinegar, salt, mustard seeds, and 2 tablespoons/30 grams of the sugar in a large saucepan and bring to a simmer, then reduce the heat and simmer gently for 10 minutes. Drain (discard the liquid), and transfer to a jar or heatproof bowl.

2. Combine the remaining sugar, celery seeds, white wine vinegar, allspice, and turmeric in a small saucepan and bring to a boil over high heat, then pour the hot mixture over the cucumbers and onions. Allow to cool to room temperature, then cover and refrigerate.

Yield: 1 quart/1 liter

ACKNOWLEDGMENTS

The authors would like to acknowledge the many people who gave us time, information, wisdom, and/or support in the writing of this book.

Foremost, Maria Guarnaschelli, senior editor at W. W. Norton, as well as Erik Johnson, Aaron Lammer, Melanie Tortoroli, Will Glovinsky, and Mitchell Kohles; copy editor Judith Sutton; and our agent, Elizabeth Kaplan.

At the Culinary Institute of America, Michael Pardus provided spirited help with the chapter on salting; Richard Virgili discussed safety issues with us. Lyde Buchtenkirk-Biscardi helped us find the origins of the techniques used in making emulsified sausages, and Fritz Sonnenschmidt was generous in discussing different methods for the emulsified sausage.

Harold McGee was unfailingly available to answer questions about the chemical and physical interactions of salt, water, and protein, as well as the secret lives of bacteria ("Can't live with 'em, can't live without 'em") and other good and bad microorganisms, and safety issues regarding nitrites, nitrates, smoking, and dry-curing, often sending us information from industry publications we'd never have located on our own. Molly Stevens gave us a last-minute read that was very helpful.

Joseph Sebranek, a meat-science expert and professor at Iowa State University, helped with nitrite, nitrate, salt, and other dry-curing questions. Brian attended that university's meat-science seminar and would like to express his gratitude to those in the meat laboratory there. Lynn Knipe, a professor at Ohio State University, was also helpful.

Armandino Batali of Salumi Artisan Cured Meats in Seattle, and Mark Buzzio, of Salumeria Biellese in Manhattan, were informative in their discussions of dry-curing.

Chef Eric Ziebold contributed to our ongoing discussions of sausage making.

Peter Kaminsky, author of *Pig Perfect*, helped with dry-cured ham information.

This book owes much to the good fortune of Brian's position as a chef-instructor in the culinary arts program at Schoolcraft College in Livonia, Michigan. He thanks the administration there for the use of their extraordinary facilities; Shawn Loving, director of culinary arts, and his colleagues, for their ideas and support; and the many students who, wittingly or unwittingly, participated in recipe development and testing. Especially valuable were his sous-chefs there, Zak Kuczynski, Adam Shulte, Justin Swain, and more recently, chef assistant Emilia Juocys.

Brian also expresses enormous gratitude to the staff at his restaurants, especially David Gilbert, former executive chef of Forest Grill in Birmingham, Michigan, who worked harder at the restaurant so that Brian could take time off to work on recipes.

And Brian would like to thank his mentor, Milos Cihelka, for getting him started.

We extend special thanks to Dan Hugelier, an instructor at Schoolcraft, a certified master chef of uncommon erudition, talent, and wisdom, for his help with this book and so much else. Some of the recipes in this book were built from the groundwork initiated and conceived by Dan in a lab book used in the charcuterie class at Schoolcraft College. Brian refined them over the years, then broke them down further for the home cook. We cannot overemphasize our gratitude and affection for Dan and his work. "Dan has been an invaluable contributor in the defining and refining of American cooking," Brian says. "I not only have the highest regard for him as a chef but also as a human being and friend who is always willing to give of himself for the betterment of the profession without asking for anything in return." The sharing of knowledge is fundamental to the advancement of the craft. As Dan himself says about working with the chefs who taught him, "If you have been privileged to learn from an accomplished craftsman, commit to humility and share all you can with others."

The published resources we relied on most, and we recommend them highly, were these:

Alan Davidson, *The Oxford Companion to Food*.

Prosper Montagne, ed., *Larousse Gastronomique*.

Harold McGee, *On Food and Cooking* (1984 and 2004).

The Culinary Institute of America, *Garde Manger: The Art and Craft of the Cold Kitchen*.

Also helpful, and fascinating, were:

Filipe Fernandez-Armesto, *Near a Thousand Tables*.

Paul Bertolli, *Cooking by Hand*.

Mark Kurlansky, *Salt: A World History* and *Cod: A Biography of a Fish That Changed the World*.

SOURCES

For more information visit ruhlman.com or stay connected on twitter @ruhlman.

Because it's so basic, so elemental, charcuterie requires few special ingredients and tools, but those it does require—curing salts, for instance, or farm-raised pork for dry-cured products—can be fundamental to the success of the recipes. You can find many sources for such items, but these are the sources we use and like.

Sausage-Making Supplies, Grinders, Stuffers

We've relied for years now on Fritz Blohm and his company Butcher & Packer for all our sausage needs and highly recommend that you do too. Virtually everything you need for the recipes in this book, from curing salts and bacterial cultures, to casings to sausage stuffers can be ordered from its website.

Butcher & Packer Supply Company
1468 Gratiot Avenue
Detroit, MI 48207
800-521-3188; 313-567-1250
www.butcherpacker.com

Weston Supply in Strongsville, Ohio, also offers a range of sausage making equipment, including grinders, stuffers, and slicers.

Weston Supply
20365 Progress Drive
Strongsville, OH 44149
800-814-4895
www.westonsupply.com

For pink salt (theirs is called Insta Cure #1), sodium nitrate (Insta Cure #2), and Fermento, as well as a broad range of sausage-making supplies:

The Sausage Maker Inc.
1500 Clinton St., Building 123
Buffalo, NY 14206
888-490-8525; 716-824-5814
www.sausagemaker.com

Smokers

For home smoking, we used a Bradley Smoker. Generally it's a very good smoker and there's nothing else out there for the price that allows you to smoke at low temperatures. The smoker can be purchased through the company's dealers in the United States and Canada. For dealers near you, see the company's web site, or call them:

Bradley Technologies Canada Inc.
1609 Derwent Way
Delta, British Columbia V3M 6K8
800-665-4188; 604-270-3646
www.bradleysmoker.com

The Big Green Egg (and other Kamado grills) is superb for hot-smoking (bacon, pastrami).

Big Green Egg
3417 Lawrenceville Highway
Atlanta, GA 30084
770-938-9394
www.biggreenegg.com

For chefs in restaurant kitchens, we recommend the Alto-Shaam smoker for about $5,000. For production smoking, Brian uses an Enviro-Pak smoker ($10,000).

Pork

Niman Ranch
1600 Harbor Bay Parkway
Suite 250
Alameda, CA 94502
866-808-0340 or 510-808-0330
www.nimanranch.com

Heritage Foods USA
Box 198, 402 Graham Ave.
Brooklyn, NY 11211
718-389-0985
www.heritagefoodsusa.com

American Livestock Breeds Conservancy
P.O. Box 477
Pittsboro, NC 27312
919-542-5704
www.albc-usa.org

Local Harvest
www.localharvest.org
Connect with local farmers.

Some of the Top Salumi Makers

Benton's Smoky Mountain Country Hams
2603 Hwy. 411 North
Madisonville, TN 37354
423-442-5003
www.bentonscountryhams2.com

Boccalone
Boccalone Salumeria
1 Ferry Building # 21
San Francisco, CA 94111
415-433-6500

Boccalone Plant
1924 International Blvd.
Oakland, CA 94606
510-261-8700
www.boccalone.com

Creminelli
310 Wright Brothers Dr.
Salt Lake City, UT 84116
801-428-1820
www.creminelli.com

Fra' Mani
1311 Eighth St.
Berkeley, CA 94710
510-526-7000
www.framani.com

La Quercia
400 Hakes Dr.
Norwalk, IA 50211
515-981-1625
http://laquercia.us

Newsom's Hams
208 East Main St.
Princeton, KY 42445
270-365-2482
www.newsomscountryham.com

Olympic Provisions
107 SE Washington St.
Portland, OR 97214
503-954-3663
www.olympicprovisions.com

Salumeria Biellese
378 8th Ave.
New York, NY 10001
212-736-7376
www.salumeriabiellese.com

Salumi
309 Third Ave. South
Seattle, WA 98104
206-621-8772
www.salumicuredmeats.com

Oyama Sausage Company
Box 126, 1689 Johnston St.
Grandville Island Public Market
Vancouver, BC V6H 3R9
Canada
604-327-7407
www.oyamasausage.ca

Duck and Duck Products

For all duck and duck products, including
legs, duck fat, and foie gras:

D'Artagnan, Inc.
280 Wilson Avenue
Newark, NJ 07105
800-327-1870
www.dartagnan.com

Hudson Valley Foie Gras
80 Brooks Road
Ferndale, NY 12734
845-292-2500
www.hudsonvalleyfoiegras.com

Maple Sugar

GloryBee Foods
120 N. Seneca
P.O. Box 2744
Eugene, OR 97402
800-456-7923
www.glorybeefoods.com

Pâté Molds

For terrine molds and pâté en croûte molds, as well as many other nifty tools:

J. B. Prince Company
36 East Thirty-First Street
New York, NY 10016
800-473-0577; 212-683-3553
www.jbprince.com

pH Meter and Paper

For a pH meter:

Paul N. Gardner Company, Inc.
316 N.E. First Street
Pompano Beach, FL 33060
800-762-2478; 954-946-9454
www.gardco.com

For pH paper:

Indigo Instruments
169 Lexington Court, Unit 1
Waterloo, Ontario N2J 4R9
877-746-4764; 519-746-4761
www.indigo.com

INDEX

Note: Page numbers in **boldface** refer to recipes; page numbers in *italics* refer to illustrations.